OXFORD MEDICAL PUBLICATIONS

Psychopharmacology of
Depression

BRITISH ASSOCIATION
FOR PSYCHOPHARMACOLOGY MONOGRAPHS

Psychopharmacology of depression

BRITISH ASSOCIATION FOR
PSYCHOPHARMACOLOGY MONOGRAPH
NO. 13

Edited by

STUART A. MONTGOMERY

Professor of Psychiatry
Academic Department of Psychiatry
St Mary's Hospital Medical School

and

TIMOTHY H. CORN

Glaxo Group Research
Greenford, Middlesex, UK

Oxford   New York   Tokyo
OXFORD UNIVERSITY PRESS
1994

Oxford University Press, Walton Street, Oxford OX2 6DP
Oxford New York Toronto
Delhi Bombay Calcutta Madras Karachi
Kuala Lumpur Singapore Hong Kong Tokyo
Nairobi Dar es Salaam Cape Town
Melbourne Auckland Madrid
and associated companies in
Berlin Ibadan

Oxford is a trade mark of Oxford University Press

Published in the United States
by Oxford University Press Inc., New York

A catalogue record for this book is available from the British Library

Library of Congress Cataloging in Publication Data
The Psychopharmacology of depression/edited by Stuart A. Montgomery
and Timothy H. Corn.
(British Association for Psychopharmacology
monograph; no. 13) (Oxford medical publications)
Includes index.
1. Depression, Mental–Pathophysiology. 2. Depression, Mental–
Chemotherapy. 3. Antidepressants–Mechanism of action.
I. Montgomery, S. A. II. Corn, Timothy H. III. Series. IV. Series:
Oxford medical publications.
[DNLM: 1. Depressive Disorder–drug therapy. 2. Depressive
Disorder–epidemiology. 3. Psychopharmacology. 4. Antidepressive
Agents–therapeutic use. W1 BR343D no. 13 1994/WM 171 P9745
1991]
RC537.P764 1994 616.85'27061–dc20 93–44896
ISBN 0 19 262278 1 (Hbk)

Typeset by Footnote Graphics, Warminster, Wiltshire
Printed in Great Britain by
Bookcraft (Avon) Ltd, Midsomer Norton, Avon

Preface

The British Association of Psychopharmacology has a tradition of bringing together experts in the field to address issues of current interest. This series of chapters by leading authorities in their subjects comes out of the Association's meeting on the Psychopharmacology of Depression.

Depression is easy to recognize, and routinely asking for the core symptoms and using a simple criterion of four or more out of eight core symptoms would double the chances of a depressed patient being identified. Antidepressants are effective, they are not addictive, and do not lose efficacy with prolonged use. A better quality of life and reduced chance of new episodes of depression is obtained if antidepressant treatment is continued in the long term. The newer antidepressants have fewer side effects, are significantly better tolerated, and therefore easier to use in the necessary full doses than the older antidepressants. Because of the substantial suicide risk in depression, which cannot be reliably predicted, safe antidepressants are to be preferred.

Depression is a common illness with a recurrent and chronic nature. It is not an easy illness to treat because those who suffer from it are frequently afraid of the illness and do not come forward for treatment. In its recent guidelines on treating depression, the British Association for Pharmacology (BAP) emphasized the need for greater attention to recognizing depressive illness, and for the education of practitioners about the special problems associated with depression.

This volume discusses a number of novel theories of depression, which partly derive from the advances in the discrimination of receptor subtypes. The progress in our understanding of neurohumoral mechanisms throws some light on which receptor subtypes are most important for depression and its related disorders, such as obsessive compulsive disorder. Novel theories on rapid cycling and on recurrent brief depression are discussed here, and these indicate that the spectrum of depression is far wider and more complex than has been thought. This volume concentrates on the newer mechanisms being explored as potential antidepressants and on the biochemical theories that have produced new treatment approaches.

London   S.A.M.
April 1994   T.H.C.

Contents

Contributors

J.A. BEARN, *Institute of Psychiatry, De Crespigny Park, Denmark Hill, London SE5 8AF, UK.*

PAUL BEBBINGTON, MRC *Social and Community Psychiatry Unit, Institute of Psychiatry, De Crespigny Park, London SE5 8AF, UK.*

I.C. CAMPBELL, *Institute of Psychiatry, De Crespigny Park, Denmark Hill, London SE5 8AF, UK.*

S.A. CHECKLEY, *Institute of Psychiatry, De Crespigny Park, Denmark Hill, London SE5 8AF, UK.*

TIMOTHY H. CORN, *Glaxo Group Research, Greenford, Middlesex UB6 0HE, UK.*

P.J. COWEN, MRC *Unit of Clinical Pharmacology and Oxford University Department of Psychiatry, Littlemore Hospital, Oxford OX4 4XN, UK.*

KARON DAWKINS, *Section on Clinical Pharmacology, Experimental Therapeutics Branch, National Institute of Mental Health, Bethesda, MD 20892, USA.*

J.F.W. DEAKIN, *Manchester University, Department of Psychiatry, Manchester Royal Infirmary, Oxford Road, Manchester M13 9WL, UK.*

KIRK DENICOFF, *Biological Psychiatry Branch, National Institute of Mental Health, Bethesda, Maryland, USA.*

CÉCILE DURLACH-MISTELI, *Department of Pharmacology, Rudolf Magnus Institute, University of Utrecht, The Netherlands.*

I.N. FERRIER, *School of Neurosciences, University of Newcastle upon Tyne, Newcastle upon Tyne NE1 4LP, UK.*

S.E. GARTSIDE, MRC *Unit of Clinical Pharmacology and Oxford University Department of Psychiatry, Littlemore Hospital, Oxford OX4 4XN, UK.*

MARK S. GEORGE, *Biological Psychiatry Branch, National Institute of Mental Health, Bethesda, MD, USA.*

FRED GROSSMAN, *Section on Clinical Pharmacology, Experimental Therapeutics Branch, National Institute of Mental Health, Bethesda, MD 20892, USA.*

TERENCE A. KETTER, *Biological Psychiatry Branch, National Institute of Mental Health, Bethesda, MD, USA.*

R. KUMAR, *Institute of Psychiatry, De Crespigny Park, Denmark Hill, London SE5 8AF, UK.*

B.E. LEONARD, *Pharmacology Department, University College, Galway, Ireland.*

GABRIELE S. LEVERICH, *Biological Psychiatry Branch, National Institute of Mental Health, Bethesda, MD, USA.*

MARKKU LINNOILA, *Division of Intramural Clinical and Biological Research, National Institute of Alcohol Abuse and Alcoholism, Bethesda, MD, USA.*

JUAN J. LÓPEZ-IBOR, JR, *Psychiatric Department, San Carlos Hospital, University Complutense, Madrid, Spain.*

HUSSEINI K. MANJI, *Section on Clinical Pharmacology, Experimental Therapeutics Branch, National Institute of Mental Health, Bethesda, MD 20892, USA.*

M.N. MARKS, *Institute of Psychiatry, De Crespigny Park, Denmark Hill, London SE5 8AF, UK.*

KIRSTIN MIKALAUSKAS, *Biological Psychiatry Branch, National Institute of Mental Health, Bethesda, MD, USA.*

STUART A. MONTGOMERY, *Academic Department of Psychiatry, St Mary's Hospital School, London, UK.*

RAYMOND J.M. NIESINK, *Department of Pharmacology, Rudolf Magnus Institute, University of Utrecht, The Netherlands.*

E.S. PAYKEL, *University of Cambridge, Department of Psychiatry, Addenbrooke's Hospital, Cambridge CB2 2QQ, UK.*

ROBERT M. POST, *Biological Psychiatry Branch, National Institute of Mental Health, Bethesda, MD, USA.*

WILLIAM Z. POTTER, *Section on Clinical Pharmacology, Experimental Therapeutics Branch, National Institute of Mental Health, Bethesda, MD 20892, USA.*

JERÓNIMO SAIZ, *Psychiatric Department, Ramón y Cajal Hospital, University of Alcalá, Madrid, Spain.*

JAN M. VAN REE, *Department of Pharmacology, Rudolf Magnus Institute, University of Utrecht, The Netherlands.*

ROSA VIÑAS, *Psychiatric Department, Ramón y Cajal Hospital, University of Alcalá, Madrid, Spain.*

MATTI VIRKKUNEN, *Department of Psychiatry, University of Helsinki, Helsinki, Finland.*

A. WIECK, *Institute of Psychiatry, De Crespigny Park, Denmark Hill, London SE5 8AF, UK.*

1

The epidemiology of depressive illness

PAUL BEBBINGTON

It is impossible to survey with justice the whole field of the epidemiology of depressive illness in a single chapter. I have therefore deliberately chosen to concentrate on three issues. The first is that of prevalence, in particular whether towards the end of the twentieth century we are approaching a state where we can feel happy comparing the results of different surveys. The second issue is whether, as suggested by some authors, the frequency of depressive illness is increasing. Finally, I will question whether there is any *epidemiological* basis for distinguishing between severe and mild forms of depressive illness.

THE PREVALENCE OF DEPRESSIVE ILLNESS

It is a truism that the prevalence of any disorder depends on how we choose to define it. The definition of depressive illness has never been arbitrary, since it always arose from the efforts of physicians to describe the phenomena that they saw in their patients. The current ideas about the severer forms of depressive illness can be traced back to their origins in nineteenth-century psychiatry, if not beyond (Bebbington 1987). They owe a lot to the experience of psychiatrists in asylums for the insane. However, even in the nineteenth century psychiatrists were building up practices with fashionable patients outside these hospitals, and were already broadening their concepts of depressive illness as a consequence (Oppenheim 1991).

Over the years we have sought to refine these ideas into some kind of a consensus. Despite these efforts, there was scant evidence of success at the time when the Dohrenwends reviewed community surveys of psychiatric disorders quarter of a century ago (Dohrenwend and Dohrenwend 1968). However, the intervening period has seen both the development and a convergence of instruments designed to establish the existence of psychiatric cases in the general population.

The first two generations of population surveys (so described by Dohrenwend and Dohrenwend 1982) either relied on the clinical judgement of a very few psychiatrists or on questionnaire measures designed to predict a general dimension of mental ill health without any diagnostic intent. The current, third, generation of psychiatric community surveys is viewed with considerably more approval, and with reason, since they rely on well standardized case finding instruments. They fall into two main groups. The first group comprises surveys based on the Present State Examination (PSE—Wing *et al.* 1974). They have been conducted largely n Europe and Australasia. The other group started with the ECA studies in the USA, but includes investigations elsewhere; their unifying feature is reliance on the Diagnostic Interview Schedule (DIS—Robins *et al.* 1985).

Although these two instruments are both standardized, they represent very different approaches to the identification of psychopathology in the general population. The development of the PSE involved an initial specification of moods, experiences, and behaviours that were regarded as abnormal, and relevant to the conditions it sought to investigate. These were then defined as a series of items and incorporated into a glossary. The items formed the basis for an interview with a structure that ensured the item field was covered and items elicited in a comparable manner by different interviewers. Because many people require an overall classification for their purposes, an explicit algorithm was developed to provide it; this was in turn embodied in the computer program CATEGO. The classes obtainable through CATEGO are based on the International Classification of Disease (ICD-9—WHO 1978).

The CATEGO program will provide a classification even in cases with very few symptoms. This poses a problem for its use in general population surveys, as case definition then becomes so broad as to be meaningless. The problem was obviated by developing a subprogram, the Index of Definition (Wing and Sturt 1978). This uses the PSE data to allocate each case to one of eight *levels* that reflect the reliance that can be placed on the CATEGO classification. Level 5 is a threshold set deliberately low, and Levels 6 to 8 are held to indicate definite cases. In population surveys, Levels 5 and above have generally been taken as the definition of a case.

A central aspect of the PSE is the manner in which it is concluded that an item should be recorded as present. This is a process of *matching*, whereby the interviewer adjudges whether accounts solicited from subjects reflect experiences that correspond to the interviewer's concepts of given items. These concepts, it is hoped, have been standardized by perusal of the glossary, and from the case law that emerges in the course of training. Interviewers are allowed to formulate questions in addition to those in the schedule in order to establish whether an item is present. Comparability is thus pursued not by rigidity of interview but by ensuring that interviewers share item concepts as exactly as possible. Good reliability is achieved with this approach. The instrument clearly depends on administration by people of some clinical sophistication, although it has been used as a screening device by lay interviewers (Wing *et al.* 1977; Sturt *et al.* 1981).

In the USA there has been a long tradition of using lay interviewers for obtaining information about mental conditions in the general population. As a result, researchers have been concerned not to rely upon clinical skills but to eliminate the need for them. The DIS deliberately incorporates a very rigid structure: it is a list of questions whose form is exactly pre-scribed. Interviewers are trained not to deviate from the printed format, so that the scope for clinical judgement is reduced to a minimum. No glossary of items is therefore required. This strategy is understandable in the context of the instrument's development, but most European psychiatrists would feel that it purchases reliability and comparability at the expense of validity. This is because the validity of diagnostic concepts is dependent on the validity of the symptoms that are the fundamental components of syndromes, and these have themselves been formalized on the basis of medical cross-examination. Thus, if you cannot be sure what the answer to a question means in terms of the concepts that inform it, 'noise' is introduced into the system.

The data obtained by the DIS can be used to establish diagnostic classes according to three different sets of criteria, of which the most important is DSM-III (APA 1980). Here I shall be particularly concerned with the DSM-III diagnosis of Major Depressive Disorder.

Let us consider the prevalence of depressive disorder as defined by CATEGO and by the Major Depressive Disorder category of DSM-III. Results are summarized in Tables 1.1 and 1.2. I have grouped the PSE results into three. The first group is Northern European and Australasian. The second comprises Mediterranean populations (although one sample was resident in London). The third group is made up of studies from the developing world.

It will be seen that for males the range of prevalences in the Northern European group is from 2.4 to 4.8 per cent, while for females it is 6.5–9.0 per cent. Although the Mediterranean group reveal higher prevalences overall, the rates for actual depressive disorders are not very different from group 1, ranging from 4.5 to 5.2 per cent for males and 7.1–11.0 per cent for females. The two highest values come from the three studies in the developing world. The variations within the groups can be explained fairly plausibly in terms of differences between the areas surveyed.

Now consider the results using the DIS. These too are reasonably consistent, provided we exclude the studies from Taiwan and from Christ-church. These two last named studies are difficult to explain. The Taiwan study provides very low prevalence estimates, in contradistinction to other studies suggesting a fairly average rate of minor psychiatric disorder in that island (for example, Cheng 1989). Likewise, the PSE study in New

TABLE 1.1. *Prevalence of CATEGO depressive classes (%)*

Site	N	Male	Female	Total
Group 1 Canberra (Henderson *et al.* 1979)	756 (157)	2.6	6.7	4.8
Camberwell (Bebbington *et al.* 1981)	800 (310)	4.8	9.0	7.0
Edinburgh (Surtees *et al.* 1983)	576		6.9	
Nijmegen (Hodiamont *et al.* 1987)	3232 (486)*			5.5
Finland† (Lehtinen *et al.* 1990)	742	2.4	6.5	4.6
Group 2 Athens (Mavreas *et al.* 1986)	489	4.3	10.1	7.4
Santander (Vazquez-Barquero *et al.* 1987)	1223 (452)*	4.5	7.8	6.2
Camberwell (Cypriot) (Mavreas and Bebbington 1987)	307	4.2	7.1	5.6
Sardinia (Carta *et al.* 1991)	374	5.2	11.0	8.3
Group 3 Uganda (Orley and Wing 1979)	206	17.0	21.0	18.9
Palembang Bahar (1989)	839 (100)*			6.5
Dubai (Ghubash *et al.* 1992)	300		13.7	

* Two-stage survey—numbers given PSE at second stage in brackets. Prevalences weighted to represent original population.

† Age adjusted figures.

Studies that do not give a breakdown in terms of CATEGO classes are not included.

TABLE 1.2. *Six-month prevalence of major depressive episode (%)*

Site	N	Male	Female	Total
ECA studies				
Baltimore (Myers *et al.* 1984)	3481	1.3	3.0	2.2
New Haven (Myers *et al.* 1984)	3058	2.2	4.6	3.5
St Louis (Myers *et al.* 1984)	3004	1.7	4.5	3.2
Piedmont* (Blazer *et al.* 1985)	3921			1.7
Los Angeles (Burnham *et al.* 1987)	3125			3.1
Other sites				
Puerto Rico (Canino *et al.* 1987)	2.4	3.3	3.0	
Edmonton (Bland *et al.* 1988)	3258	2.5	3.9	3.2
Taiwan† (Hwu *et al.* 1989)				
Taipei	5004			0.6
Small town	3005			1.1
Rural	2995			0.8
Christchurch (Oakley-Browne *et al.* 1989)	1498	3.4	7.1	5.3
Munich (Wittchen *et al.* 1992)	1366 (501)‡			3.0

* Weighted to take account of over-sampling of elderly.
† One-year prevalence.
‡ Two-stage survey—numbers given DIS at second stage in brackets. Prevalences weighted to represent original population.

Zealand by Romans-Clarkson *et al.* (1990) gives little reason to expect a very high rate of depressive disorder there. If these two studies are excluded, the range of prevalence of Major Depressive Disorder for males is 1.3–2.5 per cent, and for females 3.0–4.6 per cent. This really is quite a consistent result, and suggests that the Major Depressive Disorder category has a slightly higher threshold than depressive disorder defined by ID5 under the PSE. Note that it does not need to be a very different threshold to make quite a lot of difference to prevalence, since we know that these conditions in the community in any case tend to cluster around the threshold.

When lay interviewers use the DIS they are not very good at identifying cases confirmed as such by psychiatrists, either using the DIS itself, or an augmented clinical instrument (Anthony *et al.* 1985; Helzer *et al. 1985*). *Lay interviewers may not therefore pick out cases correctly from the general population.* However, this does not necessarily mean that the prevalences obtained are very wrong, as the distortions arising from false positives and false negatives may cancel out.

The PSE-based prevalences become much more like the DIS results if, instead of taking ID5 as the level of case definition, we take ID6. In most studies this eliminates two-thirds of identified cases.

So here we have two standardized instruments that give reasonably consistent results but prevalences that are somewhat at variance. Which are we to choose? Are there any grounds for choosing between them? Are these definitions as arbitrary as any others that might be thought of?

The major problem in considering affective disorders in the general population is that it has been clear for some time that affective symptoms are widely distributed: many people have a few symptoms, a few people have many. It could be argued that the experience of affective disturbance in the community is a seamless garment, and that we should not be attempting to impose categories on what is essentially a continuum. There is clearly some cogency to this viewpoint. However, it is also possible to justify a different line.

So, for instance, one of the features of mood disturbance that render it of interest to the psychiatric professions is that it impairs social performance. In our own study, we were able to relate impaired social performance to ID level (Hurry *et al.* 1983). When we did this, we observed that there were two points of discontinuity. one was between ID4 and ID5 and another was between ID5 and ID6. In other words, moving from ID4 to ID5, or from ID5 to ID6, causes a sudden jump in the associated degree of social impairment. This would give some justification for using either of these distinctions as the basis of case definition, slightly more for the higher threshold.

Another discontinuity concerns the effectiveness of antidepressant medication. It would appear that antidepressants are more effective in patients who meet the criteria for Major Depressive Disorder. Indeed, virtually no effect can be demonstrated in those who fall short of these criteria (Paykel *et al*. 1988).

Finally, in our study we gave our own clinical judgement about whether subjects should be regarded as cases or not. When we did this, our threshold for identifying definite cases was higher than the threshold level on the Index of Definition. This is in line with the original intention to set the threshold (ID5) at a deliberately low point. Obviously, we cannot place too much weight on this, but there is an argument for defining cases at a level that reasonably seasoned clinicians find comfortable.

Ultimately, cases are defined in the light of what is useful. This criterion may change with increasing knowledge, but at the moment there is a reasonable argument for going along with the sort of threshold that is identified by the criteria for Major Depressive Disorder. This, in turn, is likely to be largely congruent with the equivalent ICD-10 categories (WHO 1992). Using this threshold, we can reckon that the prevalence of depressive illness in UK communities with an average social profile may be around 2–3 per cent or so. It will be higher in more disadvantaged communities and groups—inner city areas and women with young children, for instance.

This threshold and this prevalence also corresponds to what is known of depressive illness behaviour in the general population. In our own study only 22 per cent of subjects at ID5 level had visited a doctor with their nerves in the last year, in contrast to three-quarters of those at ID6 and above (Hurry *et al*. 1987).

ARE WE REALLY ENTERING AN AGE OF MELANCHOLY?

However, the prevalence of depressive disorders may not be set for all time: indeed it would be surprising if it were. Several authors have raised the possibility that the frequency of depression is increasing as the century proceeds (Klerman 1978, 1988; Schwab *et al*. 1979). Klerman (1988) has recently reviewed the evidence for this, and rehearses the distinction between *age*, *period*, and *cohort* effects. (A cohort is a group with birth dates occurring within a defined period, say the 10 years from 1945 to 1954.) Subjects may be exposed to risk of disorder because they are passing through *an age of risk* (an age effect). This will be revealed in all cohorts that have passed that age. They may equally be exposed to disorder because they are passing through *a time of risk* (for instance, 1957 for the Asian 'flu' epidemic). In this instance, different cohorts will suffer the disorder at different ages, corresponding to their age at the relevant date (a period effect). Finally, cohorts may have differing lifetime risks because they are just different groups of people—the differences can be related neither to a given age nor to a given date, and there is a true *cohort effect*.

There are a number of types of study that might reveal changing incidence and prevalence of disorder. The crudest indicator would be the finding that the mean rates from a range of surveys within a geographical region were changing over time. However, if the actual geographical areas in which the studies are conducted are widely divergent, no firm conclusion can be drawn.

More weight can be placed on studies carried out at different times in a single area. However, little can be inferred from such surveys if they rely on widely differing methods of assessment; uniformity of method in population surveys is too recent to be of much use.

People have been referred to psychiatric services and admitted to mental hospitals in an apparently similar way for a considerable time, so it is possible to use referral or admission data to study the issue. However, even here the policies of service agencies may change with time and give a spurious indication of changing rates. Unfortunately, the more time needed to demonstrate change in rates, the more likely that policies will have changed.

In this respect, national admission statistics have the advantage of covering a single geographical area, but are particularly likely to reflect changes in nosocomial factors. Better data may be obtained from case registers. These are based on relatively uniform referral and diagnostic practices, although few remain in existence for more than a decade or two. Data from the Camberwell Register over the period 1964–82 actually show a *decline* in the first referral rates for depressive conditions. It is not clear whether this might represent a period or a cohort affect, or merely the consequences of diagnostic vagaries. As the age structure of the population has changed little in this time, an age effect is unlikely to account for these findings.

There have been two instances where community surveys have been repeated in the same area after several years, using the same or translatable methods. These are the Stirling County Study (Leighton *et al*. 1963; Murphy *et al*. 1984) and the Mid-town Manhattan Study (Srole *et al*. 1962; 1980). If anything, they suggest that minor psychiatric disorders are becoming less prevalent, although once again it is impossible to distinguish period and cohort effects.

There has also been a study in which *the same population* has been reinterviewed after a number of years. Hagnell *et al*. (1982) found that data collected in 1947, 1957, and 1972 in a Swedish rural community ('Lundby'), supported the suggestion that the incidence of severe depression is decreasing, but that depression of mild to moderate severity is increasing, particularly in young adult males.

Recently, data from large family studies and community surveys of affective disorder have been used to analyse the age-specific incidence for successive cohorts of subjects (Klerman *et al.* 1985; Gershon *et al.* 1987; Lavori *et al.* 1987; Joyce *et al.* 1990; Burke *et al.* 1991). To recapitulate, cohort effects imply variations in the morbid risk of depression according to date of birth. It is possible to use *lifetime prevalence* (the proportion of the population that is suffering or has suffered from a given disease) obtained through cross-sectional surveys as a first indication of changes in risk in succeeding cohorts. If there is no cohort effect, or there is a reduction in depressive experience with later birth date, cross-sectional surveys would be expected to show an increase in life time prevalence in older subjects, as they have had a longer life in which to get depressed.

In fact, there is a consistent *decline* in lifetime prevalence with age in a number of studies, including our own (Bebbington *et al.* 1989) (see Fig. 1.1). This suggests a cohort effect of increasing, and possibly earlier, incidence of depression in subjects with later birth dates. The direct examination of this possibility appears to confirm it. Klerman (1988) has reviewed the available databases to illustrate that the relationship between age and incidence of depression has become steeper in succeeding birth cohorts (see Fig. 1.2 and 1.3, redrawn from Klerman *et al.* 1985). In other words, as the twentieth century proceeds, the population is becoming increasingly prone to depression, and this emerges at a younger age. In public health terms, this is a potential disaster of enormous proportions, as Kramer *et al.* (1987) have pointed out.

These illustrations are persuasive, but depend crucially on the subject's ability to remember past episodes. Such episodes are likely to be more

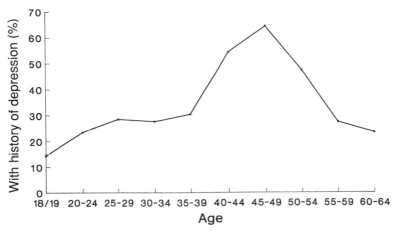

FIG. 1.1. Relationship between lifetime history of depression and age.

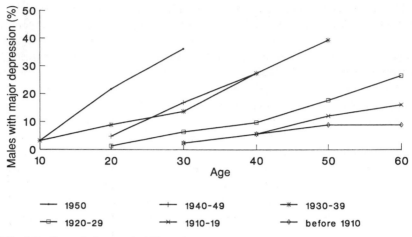

FIG. 1.2. Cumulative probability of major depression by cohort.

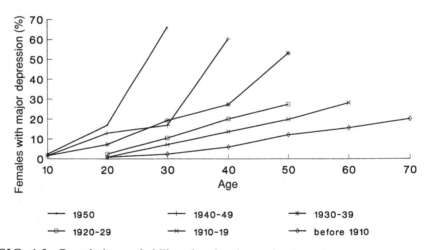

FIG. 1.3. Cumulative probability of major depression by cohort.

remote, and thus easier to forget, in older subjects, and this may account for the findings described by Klerman. Estimates of lifetime prevalence do seem to be of doubtful validity (Bromet *et al.* 1986). Recently, Hasin and Link (1988) have also raised the possibility that older subjects are less able to recognize past disturbances in their functioning as psychological in nature—this would also be a cohort effect, but of a different type.

We must therefore reserve judgement. Nevertheless, the possibility of increasing proneness to depression in sequential birth cohorts merits very serious attention, particularly in view of its public health implications.

USING EPIDEMIOLOGICAL FINDINGS TO VALIDATE THE DISTINCTION BETWEEN MILD AND SEVERE DEPRESSION

The most consistent attempts to discriminate between mild and severe depressive illness had arisen from the concept of endogeneity. This has a number of meanings (Lewis 1972), one of which is that affective disturbances with more extreme symptoms are unlikely to be responses to unfortunate events or circumstances. The theme of endogeneity has been so prominent that certain symptoms are now referred to as *endogenous* whether they were provoked by adversity or not. Nevertheless, it has been very difficult to demonstrate that there is much difference between severe and mild depressions in terms of the likelihood that they have been precipitated by social misfortunes and hardships (Bebbington 1991).

Another possible distinction is that the female-to-male sex ratio for depressive disorders is less for the more severe disorders. Although the recent and enormous ECA studies from the USA display a ratio on the low side (Regier *et al.* 1988), the ratio does tend to be higher in the community than in patient series (Bebbington 1988). Moreover, in all referred series of patients where results are quoted separately for affective psychosis and for depressive neurosis, the ratio is greater in the latter (for example, Baldwin 1971; Weeke *et al.* 1975; Der and Bebbington, 1987). The effect is consistent but not major. One possible explanation is that part of the female preponderance is accounted for by social rather than biological factors, these operating most noticeably on the milder disorders and therefore most noticeably in the general population.

A line of investigation with more striking results concerns personality. Kendell and di Scipio (1968) found that patients with 'endogenous' depressions were less neurotic and less introverted following recovery than patients with 'neurotic' depressions. Workers reporting similar results include Kerr *et al.* (1970), Perris (1971), Paykel *et al.* (1976), and Benjaminsen (1981). Patients with bipolar depression are especially unlikely to show introverted and neurotic traits (Hirschfeld and Klerman 1978), and personality disorders are apparently much less common in melancholia than in less severe depressive disorders (Charney *et al.* 1981). All these authors agree that 'nonendogenous' depression is frequently precipitated by adverse environmental circumstances operating on someone whose personality is already somehow vulnerable. Parker *et al.* (1987) claim that the association of the experience of poor parenting is apparent in neurotic depressives, but not in endogenous forms of the disorder, and this might be related to the identified personality differences.

However, there is one consistent difference between severe and mild depressive illnesses that may be important. This is the relationship of incidence to age. It has long been observed that the peak age of first admissions for depressive psychosis is in late middle or early old age (Shepherd 1957; Norris 1959; Jaco 1960; Silverman 1968). I have recently reviewed both national data relating to first admissions, and register data for first inceptions (Bebbington 1988). Examples of these findings, taken from Der and Bebbington (1987), are given in Figs 1.4 and 1.5. The general conclusion is that the incidence of severe depressive illness increases with age, while for milder depressions of the type often referred to as neurotic, the peak age of incidence is much younger, usually in the thirties.

This situation might arise because older people with newly arising neurotic depressive disorder just do not get to see psychiatrists. This was shown by Kessel and Shepherd (1962). However, the findings of community surveys suggest that the effect is not of major significance. Thus, in surveys of people under sixty-five, the oldest age group is not that with the highest prevalence (Bebbington *et al.* 1981; Surtees *et al.* 1983; Myers *et al.* 1984), and when elderly folk are directly compared with their younger counterparts in the general population using the same instruments they show if anything a lower prevalence of affective disorders (Blazer *et al.* 1986). In addition, Weissman and Myers (1978) report that, once depression has developed, age has no effect on the likelihood that people will contact services.

These community findings argue against the idea that severe disorder increases in incidence with age because of the adverse social situation of the elderly in many areas of the developed world. If the increase were due to adversity, a corresponding increase in the milder types would also be expected: if the age distribution of severe disorders is the result purely of adverse social influences, these should affect the incidence of minor disorder similarly. We have seen that this is not the case.

Another interpretation is that age may act to increase the severity of disorders brought about by psychosocial adversity. However, there are two reasons why this is unlikely. If it were true, one would expect the *total incidence* of depression of all degrees of severity to be maintained in the elderly, but in fact it falls (Figs. 1.4 and 1.5). Moreover, the age–incidence curve for severe depression is actually *reversed* in those who probably have most experience of loss and adversity, that is, the widowed and divorced. This also tends to discount an effect of adversity on severity: it appears almost as if age protects against the effects of these losses, which after all probably imply different things to the elderly.

One further possibility is that the differential effect of age on severe and mild depressions is a cohort phenomena of the type we have discussed above. This could be explained psychologically in terms of a 'kettle lid model'. Thus, as suggested by Hasin and Link (1989), our parents and grandparents may have been less inclined towards psychological explana-

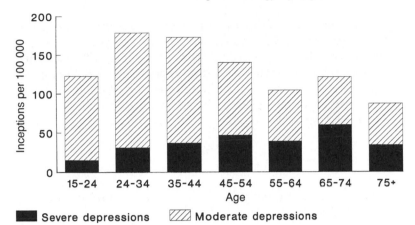

FIG. 1.4. Males—age-specific incidence of affective disorders (Camberwell).

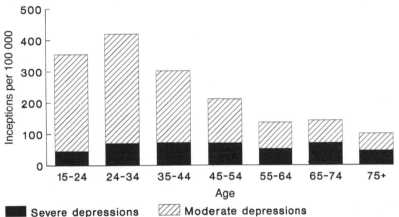

FIG. 1.5. Females—age-specific incidence of affective disorders (Camberwell).

tions for their discomfort, and possibly also towards exploration of their own emotional states. This particular coping strategy might ensure a relative immunity to moderate depression, but at the cost of increasing the risk of more florid illness. To pursue the analogy, the lid is held on until it is blown off violently by the increasing head of steam. Our generation, in contrast, may express minor states of dysphoria more easily and so prevent severe depression building up. The longitudinal Lundby Survey can be interpreted in this light, at least in males (Hagnell *et al.* 1982). However, the age effects appear in studies separated by quite considerable periods (see Norris 1959; Der and Bebbington 1987), and this may indicate that the effect is genuinely age-related, rather than a cohort effect.

My own interpretation is that the age effect in moderate depression arises because of social influences, but that ageing exerts a *releasing* effect on severe depression. The age effect in severe depression is open to reversal, but only by social circumstances of the most adverse character, such as divorce and widowhood. It thus appears like a default option. It probably has a biological basis, although the contribution of the neuro-biology of ageing to the susceptibility to depression remains ambiguous (Veith and Raskind 1988).

CONCLUDING REMARKS

In this chapter, I have offered a personal view of current themes in the epidemiological study of affective disorder. By doing so, I hope I have convinced the reader that the practice of epidemiology is capable of providing important insights that would otherwise be unavailable.

REFERENCES

American Psychiatric Association (1980). Diagnostic and statistical manual of mental disorders (DSM III), (3rd edn). APA, Washington DC.

Anthony, J.C., Folstein, M., Romanoski, A.J., Von Korff, M.R., Nestadt, G.R., Chahal, R. *et al.* (1985). Comparison of the lay Diagnostic Interview Schedule and a standardized psychiatric diagnosis. *Archives of General Psychiatry*, **42**, 667–76.

Bahar, E. (1989). Development and mental health. A psychiatric epidemiological study in a developing country: an enquiry in Palembang, Indonesia. Ph.D. thesis. Australian National University.

Baldwin, J.A. (1971). Five year incidence of reported psychiatric disorder. In *Aspects of the epidemiology of mental illness: studies in record linkage* (ed. J.A. Baldwin). Little Brown, Boston.

Bebbington, P.E. (1987). Misery and beyond: the pursuit of disease theories of depression. *International Journal of Social Psychiatry*, **33**, 13–20.

Bebbington, P.E. (1988). The social epidemiology of clinical depression. In *Handbook of studies in social psychiatry* (ed. A.S. Henderson and G. Burrows). Elsevier, Amsterdam.

Bebbington, P.E. (ed.) (1991). The epidemiology of affective disorders. In *Social psychiatry: theory, methodology and practice*. Transaction, New Brunswick, NJ.

Bebbington, P., Hurry, J., Tennant, C., Sturt, E., and Wing, J.K. (1981). The epidemiology of mental disorders in Camberwell. *Psychological Medicine*, **11**, 561–80.

Bebbington, P.E., Katz, R., McGuffin, P., Tennant, C., and Hurry, J. (1989). The risk of minor depression before age 65: results from a community survey. *Psychological Medicine*, **19**, 393–400.

Benjaminsen, S. (1981). Primary non-endogenous depression and features attributed to reactive depression. *Journal of Affective Disorders*, **3**, 245–59.

Bland, R.C., Newman, S.C., and Orn, H. (1988). Epidemiology of psychiatric disorders in Edmonton. *Acta Psychiatrica Scandinavica*, **77** (Suppl. 338).

Blazer, D., George, L.K., Landerman, R., Pennybacker, M., Melville, M.L., Woodbury, M., Manton, K.G., Jordan, K. and Locke, B. (1985). Psychiatric disorders: a rural/urban comparison. *Archives of General Psychiatry*, **42**, 651–6.

Bromet, E.J., Dunn, L.O., Connell, M.O., Dew, M.A., and Schulberg, H.C. (1986). Long-term reliability of diagnosing lifetime major depression in a community sample. *Archives of General Psychiatry*, **43**, 435–40.

Burke, K.C., Burke, J.D., Rae, D.S., and Regier, D.A. (1991). Comparing age at onset of Major Depression and other psychiatric disorders by birth cohorts in five US community populations. *Archives of General Psychiatry*, **48**, 789–95.

Burnham, M.A., Hough, R.L., Escobar, J.I., Karno, M., Timbers, D.M., Telles, C.A. *et al.* (1987). Six month prevalence of specific psychiatric disorders among Mexican Americans and non-Hispanic whites in Los Angeles. *Archives of General Psychiatry*, **44**, 687–94.

Canino, G.J., Bird, H.R., Shrout, P.E., Rubio-Stipec, M., Bravo, M., Martinez, R. *et al.* (1987). The prevalence of specific psychiatric disorder in Puerto Rico. *Archives of General Psychiatry*, **44**, 727–35.

Carta, M.G., Carpiniello, B., Morosini, P.L., and Rudas, N. (1991). Prevalence of mental disorders in Sardinia: a community study in an inland mining district. *Psychological Medicine*, **21**, 1061–71.

Charney, D.S., Nelson, J.C., and Quinlan, D.M. (1981). Personality traits and disorder in depression. *American Journal of Psychiatry*, **138**, 1601–4.

Cheng, T.A. (1989). Urbanisation and minor psychiatric morbidity. A community study in Taiwan. *Social Psychiatry and Psychiatric Epidemiology*, **24**, 309–16.

Der, G. and Bebbington, P.E. (1987). Depression in inner London: a register study. *Social Psychiatry*, **22**, 73–84.

Dohrenwend, B.P. and Dohrenwend, B.S. (1969). *Social status and psychological disorder: a causal inquiry*. Wiley, New York.

Dohrenwend, B.P. and Dohrenwend, B.S. (1982). Perspectives on the past and future of psychiatric epidemiology. *American Journal of Public Health*, **72**, 1271–9.

Gershon, E., Hamovit, J.H., Guroff, J.J., and Nurnberger, J.I. (1987). Birth cohort changes in manic and depressive disorders in relatives of bipolar and schizoaffective patients. *Archives of General Psychiatry*, **44**, 314–19.

Ghubash, R., Hander, E., and Bebbington, P.E. (1992). The Dubai Community Psychiatric Survey: prevalence and sociodemographic correlates. *Social Psychiatry and Psychiatric Epidemiology*, **27**, 53–61.

Hagnell, O., Lanke, J., Rorsman, B., and Öjesjö L. (1982). Are we entering an age of melancholy? Depressive illness in a prospective epidemiological study over 25 years: the Lundby Study, Sweden. *Psychological Medicine*, **12**, 279–89.

Hasin, D. and Link, B. (1988). Age and recognition of depression: implications for a cohort effect in major depression. *Psychological Medicine*, **18**, 683–8.

Helzer, J.E., Robins, L.N., McEvoy, L.T., Spitznagel, E.L., Stoltzman, R.K., Farmer, A., and Brockington, I.F. (1985). A comparison of clinical and diagnostic interview schedule diagnoses. Physician reexamination of lay-interviewed cases in the general population. *Archives of General Psychiatry*, **42**, 657–66.

Henderson, A.S., Byrne, D.G., and Duncan-Jones, P. (1981). *Neurosis and the social environment*. Academic Press, Sydney.

Hirschfeld, R.M.A. and Klerman, G.L. (1979). Personality attributes and affective disorders. *American Journal of Psychiatry*, **136**, 67–70.

Hodiamont, P., Peer, N., and Syben, N. (1986). Epidemiological aspects of psychiatric disorder in a Dutch Health Area. *Psychological Medicine*, **16**, 495–506.

Hurry, J., Sturt, E., Bebbington, P., and Tennant, C. (1983). Sociodemographic association with social disablement in a community sample. *Social Psychiatry*, **18**, 113–21.

Hurry, J., Bebbington, P.E., and Tennant, C. (1987). Psychiatric symptoms and social disablement as determinants of illness behaviour. *Australian and New Zealand Journal of Psychiatry*, **21**, 68–74.

Hwu, H.-G., Yeh, E.-K., and Chang, L.-Y. (1989). Prevalence of psychiatric disorders in Taiwan defined by the Chinese Diagnostic Interview Schedule. *Acta Psychiatrica Scandinavica*, **79**, 136–47.

Jaco, E.G. (1960). *The Social Epidemiology of Mental Disorders*, Russell Sage Foundation, New York.

Joyce, P.R., Oakley-Browne, M.A., Wells, J.E., Bushnell, J.A., and Hornblow, A.R. (1990). Birth cohort trends in major depression: increasing rates and earlier onset in New Zealand. *Journal of Affective Disorders*, **18**, 83–9.

Kendell, R.E. and di Scipio, W.J. (1968). Eysenck Personality Scores of patients with depressive illnesses. *British Journal of Psychiatry*, **114**, 767–70.

Kerr, T.A., Shapira, K., Roth, M. and Garside, R.F. (1970). The relationship between the Maudsley Personality Inventory and the course of affective disorders. *British Journal of Psychiatry*, **116**, 11–19.

Kessel, N. and Shepherd, M. (1962). Neurosis in hospital and general practice. *Journal of Mental Science*, **108**, 159–66.

Klerman, G.L. (1978). Affective disorders: In *The Harvard guide to modern psychiatry* (ed. M. Armand and M.D. Nicholi). Belknap Press, Cambridge, MA.

Klerman, G.L. (1988). The current age of youthful melancholia: evidence for increase in depression among adolescents and young adults. *British Journal of Psychiatry*, **152**, 4–14.

Klerman, G.L., Lavori, P.W., Rice, J., Reich, T., Endicott, J., Andreason, N.C. *et al.* (1985). Birth-cohort trends in rates of major depressive disorder among relatives of patients with affective disorder. *Archives of General Psychiatry*, **421**, 689–93.

Kramer, M., Brown, H., Skinner, M., Anthony, J., and German, P. (1987). Changing living arrangements in the population and their potential effect on the prevalence of mental disorder: findings of the East Baltimore mental health survey. In *Psychiatric epidemiology: progress and prospects* (ed. B. Cooper). Croom Helm, London.

Lavori, P.W., Klerman, G.L., Keller, M.B., Reich, T., Rice, J., and Endicott, J. (1987). Age–period–cohort analysis of secular trends in onset of major depression: Findings in siblings of patients with major affective disorder. *Journal of Psychiatric Research*, **21**, 23–35.

Lehtinen, V., Lindholm, T., Veijola, J., and Vaisanen, E. (1990). The prevalence of PSE-CATEGO disorders in a Finnish adult population cohort. *Social Psychiatry and Psychiatric Epidemiology*, **25**, 187–92.

Leighton, D.C., Harding, J.S., Macklin, D.B., MacMillan, A.M., and Leighton, A.H. (1963). *The character of danger (Stirling County study vol. 3)*. Basic Books, New York.

Lewis, A. (1971). 'Endogenous' and 'exogenous': a useful dichotomy? *Psychological Medicine*, **1**, 191–6.

Mavreas, V. and Bebbington, P.E. (1987). Psychiatric morbidity in London's Greek Cypriot community. I. Association with sociodemographic variables. *Social Psychiatry*, **22**, 150–9.

Mavreas, V.G., Beis, A., Mouyias, A., Rigoni, F., and Lyketsos, G.C. (1986). Prevalence of psychiatric disorder in Athens: a community study. *Social Psychiatry*, **21**, 172–81.

Murphy, J.M., Sobol, A.M., Neff, R.K., Olivier, D.C., and Leighton, A.H. (1984). The stability of prevalence: depression and anxiety disorders. *Archives of General Psychiatry*, **41**, 990–7.

Myers, J.K., Weissman, M.M., Tischler, G.L., Holzer, C.E., Leaf, P.J., Orvaschel, H., Anthony, J.C. (1984). Six month prevalence of psychiatric disorders in three communities: 1980–1982. *Archives of General Psychiatry*, **41**, 959–67.

Norris, V. (1959). *Mental Illness in London, Maudsley Monograph No. 6*. Oxford University Press.

Oakley-Browne, M.A., Joyce, P.R., Wells, J.E., Bushnell, J.A., and Hornblow, A.R. (1989). Christchurch Psychiatric Epidemiology Study, Part I: six month and other period prevalences of specific psychiatric disorders. *Australian and New Zealand Journal of Psychiatry*, **23**, 327–40.

Oppenheim, J. (1991). *Shattered nerves: doctors, patients and depression in Victorian England*. Oxford University Press.

Orley, J. and Wing, J.K. (1979). Psychiatric disorders in two African villages. *Archives of General Psychiatry*, **36**, 513–20.

Parker, G., Kilom, L., and Mayward, L. (1987). Parental representations of neurotic and endogenous depressives. *Journal of Affective Disorders*, **13**, 75–82.

Paykel, E.S., Klerman, G.L., and Prusoff, B.A. (1976). Personality and symptom pattern in depression. *British Journal of Psychiatry*, **129**, 327–34.

Paykel, E.S., Hollyman, J.A., Freeling, P., and Sedgwick, P. (1988). Prediction of therapeutic benefit from amitriptyline in mild depression: a general practice placebo-controlled trial. *Journal of Affective Disorders*, **14**, 83–95.

Perris, C. (1971). Personality patterns in patients with affective disorders. *Acta Psychiatrica Scandinavica*, **47** (Suppl. 221), 43–51.

Regier, D.A., Boyd, J.H., Burke, J.D., Rae, D.S., Myers, J.K., Kramer, M. *et al.* (1988). One month prevalence of mental disorders in the United States. *Archives of General Psychiatry*, **45**, 977–86.

Robins, L.N., Helzer, J.E., Orvaschel, H., Anthony, J.C., Blazer, D.G., Burnham, A. *et al.* (1985*a*). The Diagnostic Interview Schedule. In *Epidemiologic field methods in psychiatry: the NIMH epidemiologic catchment area program* (ed. W.W. Eaton and L.G. Kessler). Academic Press, Orlando, FL.

Romans-Clarkson, S.E., Walton, V.A., Herbison, G.P., and Mullen, P.E. (1988). Marriage, motherhood and psychiatric morbidity in New Zealand. *Psychological Medicine*, **18**, 983–90.

Schwab, J.J., Bell, R.A., Warheit, G.J., and Schwab, R.B. (1979). *Social order and mental health: the Florida health study*. Brunner/Mazel, New York.

Shepherd, M. (1957) *A Study of the major psychoses in an English county, Maudsley Monograph No.3*. Oxford University Press.

Silverman, C. (1968). *Epidemiology of depression*. Johns Hopkins Press, Baltimore.

Srole, L. and Fischer, A.K. (1980). The Midtown Manhattan longitudinal study vs the 'mental paradise lost' doctrine: a controversy joined. *Archives of General Psychiatry*, **37**, 209–21.

Srole, L., Langner, T., Michael, S.T., Opler, M.K., and Rennie, T.A.C. (1962). *Mental health in the metropolis*. McGraw-Hill, New York.

Sturt, E., Bebbington, P.E., Hurry, J., and Tennant, C. (1981). The Present State Examination used by interviewers from a Survey Agency: Report from the Camberwell Community Survey. *Psychological Medicine*, **11**, 185–192.

Surtees, P.G., Dean, C., Ingham, J.G., Kreitman, N.B., Miller, P.McC., and Sashidharan, S.P. (1983). Psychiatric disorder in women from an Edinburgh community: associations with demographic factors. *British Journal of Psychiatry*, **142**, 238–46.

Vazquez-Barquero, J.-L., Diez-Manrique, J.F., Pena, C., Aldana, J., Samaniego-Rodriguez, C., Menendez-Arango, J. *et al.* (1987). A community mental health survey in Cantabria: a general description of morbidity. *Psychological Medicine*, **17**, 227–42.

Veith, R.C. and Raskind, M.A. (1988). The neurobiology of aging: does it predispose to depression? *Neurobiology of Aging*, **9**, 101–18.

Weeke, A., Bille, M., Videbeck, T., Dupont, A., and Juel-Nielsen, N. (1975). Incidence of depressive syndromes in a Danish county. *Acta Psychiatrica Scandinavica*, **51**, 28–41.

Weissman, M.M. and Myers, J. (1978). Rates and risks of depressive symptoms in a United States urban community. *Acta Psychiatrica Scandinavica*, **57**, 219–31.

Wing, J.K. and Sturt, E. (1978). *The PSE-ID-CATEGO system: a supplementary manual*. Institute of Psychiatry, London.

Wing, J.K., Cooper, J.E., and Sartorius, N. (1974). *The measurement and classification of psychiatric symptoms*. Cambridge University Press.

Wing, J.K., Henderson, A.S., and Winckle, M. (1977). The rating of symptoms by a psychiatrist and non-psychiatrist: a study of patients referred from general practice. *Psychological Medicine*, **7**, 713–15.

Wittchen, H.-U., Essau, C.A., von Zerssen, D., Krieg, J.-C., and Zaudig, M. (1992). Lifetime and six-month prevalence of mental disorders in the Munich follow-up study. *European Archives of Psychiatry and Clinical Neuroscience*, **241**, 247–58.

WHO (World Health Organization) (1978). *Mental disorders: glossary and guide to their classification in accordance with the ninth revision of the international classification of diseases*. WHO, Geneva.

WHO (World Health Organization) (1992). *Tenth revision of the international classification of diseases*. WHO, Geneva.

2

Second generation antidepressants: chemical diversity but unity of action?

B.E. LEONARD

INTRODUCTION

The accidental discovery of imipramine and iproniazid as antidepressants some 30 years ago, and the subsequent finding that these drugs caused changes in the functional activity of biogenic amines in the rat brain, served to focus research onto noradrenaline and 5-hydroxytryptamine (5-HT) as a possible basis for the mode of action of antidepressants (Ayd and Blackwell 1970). At this time, the antihypertensive alkaloid reserpine was shown to cause depression in a substantial minority of patients. This effect was attributed to the ability of the drug to deplete central neuronal stores of biogenic amines (Everett and Toman 1959). Thus by the 1960s, a number of drugs had been discovered that could either alleviate or cause depression in man. Detailed studies in rodents on the mechanism of action of these drugs showed that they either prevented the intraneuronal degradation of biogenic amine neurotransmitters (the monoamine oxidase inhibitors), reduced the rate of removal of the transmitters from the synaptic cleft by impeding the reuptake (the tricyclic antidepressants), facilitated the reuptake of the transmitter thereby reducing the duration of action of the transmitter on the postsynaptic receptor (lithium salts), or enhanced the intraneuronal catablism of the biogenic amine transmitters by decreasing their storage in synaptic vesicles (reserpine). The results of such studies formed the basis for the biogenic amine theory of depression.

In the last 20 years, antidepressants have been developed that show a high degree of selectivity for the noradrenergic, serotonergic, or dopaminergic systems. In addition, other drugs with antidepressant properties have been developed that show selectivity for the GABAergic system, that modulate benzodiazepine receptors or that act as partial agonists on a subpopulation of 5-HT receptors. Table 2.1 identifies some of the antidepressants that have been developed in recent years.

While there is no evidence that the second generation antidepressants are therapeutically more effective than the 'classical' antidepressants such

TABLE 2.1. *Some second generation antidepressants*

Selective monoamine oxidase A inhibitors	Brofaromine, cimoxatone, moclobemide
α_2-adrenoceptor antagonists	Mianserin, mepirzapine, idazoxan*
5-HT reuptake inhibitors	Zimelidine, fluoxetine, fluvoxamine, citalopram, sertraline, paroxetine
Noradrenaline reuptake inhibitors	Nomifensin, †oxaprotiline, levoprotiline*
5-HT reuptake enhancers	Tianeptine
5-HT_{1A} partial agonists	Ipsapirone,* gepirone*
Benzodiazepine analogues	Alprazolam, adinazolam, zometapine*
Drugs modulating dopaminergic function	Bupropion
GABAmimetics	Fengabine,* progabide*
Novel tricyclic antidepressants	Lofepramine

† Drugs withdrawn because of serious adverse side effects.
* Clinical efficacy uncertain.

as imipramine or phenelzine, there is evidence from those that have been widely used (for example, mianserin, fluoxetine, and lofepramine) that they are less cardiotoxic and better tolerated by the patient. The question therefore arises whether antidepressants have a common mode of action that is not directly related to presumed specificity of action on a neurotransmitter in the brain.

CHANGES IN NEUROTRANSMITTER RECEPTORS IN THE MAMMALIAN BRAIN FOLLOWING CHRONIC ANTIDEPRESSANT TREATMENT

Antidepressant therapy is usually associated with a delay of 2–3 weeks before the onset of a beneficial effect (Oswald *et al.* 1972). Much of the improvement seen early in the treatment with antidepressants is probably associated with a reduction in anxiety that often occurs in the depressed patient and improvement in sleep caused by the sedative action of many of these drugs. The delay in the onset of the therapeutic response cannot be easily explained by the pharmacokinetic profile of the drugs as peak plasma (and therefore brain) concentrations are usually reached in 7–10 days. Furthermore, the 2–3 week delay is also seen in patients given ECT. Thus the amine hypothesis which was based on the acute effects of antidepressants on the reuptake of biogenic amines, or on the inhibition of monoamine

TABLE 2.2. *Changes in neurotransmitter receptors that occur in the cortex of rat brain following chronic antidepressant treatment*

Cortical β adrenoceptors	Decreased functional activity and density
Cortical α_1 adrenoceptors	Increased density
Cortical α_2 autoreceptors	Decreased functional activity
Dopamine autoreceptors	Decreased functional activity
Cortical GABA B receptors	Decreased density
Limbic 5-HT$_{1A}$ receptors	Increased density and decreased functional activity
Cortical 5-HT$_2$ receptors	Decreased density and functional activity

Data summarzed from Serra *et al.* (1979), Blier *et al.* (1988), Butler and Leonard (1988), Charney *et al.* (1990), Heal (1990).

oxidase activity, has undergone drastic revision in recent years to account for disparity between the immediate effects of antidepressants and the onset of their clinical effects. Researchers have therefore switched their attention from the actions of antidepressants on presynaptic mechanism that govern neurotransmitter synthesis, reuptake and metabolism to the changes in receptor function. Table 2.2 summarizes some of the changes in neurotransmitter receptors that occur in the cortex of rat brain following chronic antidepressant (and ECT) treatments.

In addition to these changes in GABA B and biogenic amine receptors, recent evidence by Earley *et al.* (1990) has shown that a decrease in cortical muscarinic receptors occurs in the bulbectomized rat model of depression but, like most of the changes in biogenic amine receptors, returns to control values following treatment with both typical and atypical antidepressants. Thus irrespective of the specificity of the antidepressants following their acute administration, it would appear that a common feature of all of these drugs is to correct the abnormality in neurotransmitter receptor function. Such an effect of chronic antidepressant treatment largely parallels the time of onset of the therapeutic response and forms the basis of the receptor sensitivity hypothesis of depression and the common mode of action of antidepressants.

The mechanism whereby chronic antidepressant treatment causes changes in such diverse neurotransmitter receptors is unknown. Considerable attention has recently been focused on the interaction between serotonergic and beta-adrenergic receptors which may be of particular relevance to our understanding of the therapeutic effect of antidepressants. Thus the chronic administration of antidepressants results in an enhanced inhibitory response of forebrain neurons to microiontophoretically applied 5-HT (Aghajanian and de Montigny 1978). This enhanced response is blocked by lesions of the noradrenergic projections to the cortex (Chaput and de Montigny 1988). Such an effect could help to explain the enhanced serotonergic function that arises following chronic antidepressants or ECT. Thus most antidepressants decrease the functional activity of the beta adrenoceptors that normally have an inhibitory role on the cortical 5-HT system. This results in a disinhibition or facilitation of serotonergic (and dopaminergic) mediated behaviours.

Conversely, impairment of the serotonergic system by means of selective neurotoxins or 5-HT synthesis inhibitors largely prevents the decrease in functional activity of cortical beta adrenoceptors that usually arises following chronic antidepressant treatment (Brunello et al. 1985). Recent evidence suggests that these apparent changes in beta adrenoceptors that occur following chronic antidepressant treatment are more likely to be due to an increase in the density of 5-HT_{1B} receptor sites which are also labelled by the beta adrenoceptor antagonist dihydroalprenolol, a ligand conventionally used to identify beta adrenoceptors (Stockmeier and Kellar 1989).

Both clinical and experimental studies have provided evidence that 5-HT can regulate dopamine turnover. Thus several investigators have shown that a positive correlation exists in depressed patients between the homovanillic acid and 5-hydroxyindole acetic acid concentrations in the CSF (Agren et al. 1986). In experimental studies, stimulation of the 5-HT cell bodies in the median raphé causes reduced firing of the substantia nigra where dopamine is the main neurotransmitter (Dray et al. 1976). Agren et al. (1986) have provided convincing evidence that 5-HT plays an important role in modulating dopaminergic function in many regions of the brain, including the mesolimbic system. Such findings imply that the effects of antidepressants that show an apparent selectivity for the serotonergic system could be equally ascribed to a change in dopaminergic function in mesolimic and mesocortical regions of the brain.

The recent advances in molecular neurobiology has demonstrated how information is passed from the neurotransmitter receptors on the outer side of the neuronal membrane to the secondary messenger system on the inside. The coupling of this receptor to the secondary messenger is brought about by a member of the G protein family. Beta adrenoceptors are linked to adenylate cyclase and, depending on the subtype of receptors, 5-HT is linked to either adenylate cyclase (5-HT_{1A}, 5-HT_{1B}) or phospholipase (5-HT_{2C}, 5-HT_2). Activation of phospholipase results in an intracellular increase in the secondary messengers diacylglycerol and inositol triphosphate (IP3), the IP3 then mobilizing intraneuronal calcium (see Berridge 1987). The net result of the activation of the secondary messenger systems is the increase in activity of various protein kinases that phosphorylate membrane-bound proteins to produce a physiological response (Nestler et al. 1984).

Racagni and co-workers (1991) have investigated the effect of chronic antidepressant treatments on the phosphorylation of proteins associated with the cytoskeletal structure of the nerve cell. These investigators suggest that antidepressants could affect the function of the cytoskeleton by changing the components of the associated protein phosphorylation system. In support of their hypothesis, these researchers showed that both typical (for example desipramine) and atypical (for example (+) oxaprotiline, a specified noradrenaline re-uptake inhibitor and fluoxetine, a selective HT uptake inhibitor) antidepressants increased the synthesis of a microtubule fraction possibly by affecting the regulatory subunit of protein kinase type II. These changes in cytoskeletal protein synthesis only occurred after chronic antidepressant treatments and suggest that antidepressants, beside their well established effects on receptors and transport systems, might change neuronal signal transduction processes distal to the receptor.

Another possible mechanism whereby antidepressants may change the physical relationship between neurons in the brain is by inhibiting neurite outgrowths from nerve cells. Thus Wong and co-workers (1991) showed that the tricyclic antidepressants amitriptyline, at therapeutically relevant concentrations, inhibited neurite outgrowth from chick embryonic cerebral explants *in vitro*. These investigators postulate that the reduction in the rate of synthesis of cyclic AMP caused by amitriptyline is responsible for the reduced neurite growth. While the relevance of such findings to the therapeutic effects of amitriptyline in man is unclear, they do suggest that a common mode of action of all antidepressants could be to modify the actual structure of nerve cells and possibly eliminate inappropriate synaptic contacts that are responsible for behavioural and psychological changes associated with depression.

Hypercortisolaemia occurs commonly in depression and the relative insensitivity of the pituitary–adrenal axis to the inibitory feedback effect of the glucocorticoid dexamethasone which is used as biochemical marker of the disease. One possible explanation for this insensitivity to dexamethasone may lie in the hypersensitivity of central glucocorticoid receptors.

Interest in the possible association of central glucocorticoid receptors with altered central neurotransmitter function arose from the observation that glucocorticord receptors have been identified in the nuclei of catecholamine and 5-HT containing cell bodies in the brain (Harfstrand *et al.* 1986). Burnstein and Cidlowski (1989) have shown that glucocorticoid receptors function as DNA binding proteins which can modify the transcription of genes. Chronic imipramine treatment causes an increase in glucocorticoid receptor immunoreactivity in rat brain, the changes being particularly pronounced in the noradrenergic and serotonergic cell body regions (Kitayama *et al.* 1988). Preliminary clinical studies show that lymphocyte glucocorticoid receptors are abnormal in depressed patients (Giller *et al.* 1991). Such findings lend support to the hypothesis that the changes in central neurotransmission occurring in depression are a reflection of the effects of chronic glucocorticoids on the transcription of proteins which play a crucial role in neuronal structure and function. Sulser and Sanders-Bush (1987) have formulated a 5-HT–noradrenaline–glucocorticoid link hypothesis of affective disorders based on such findings. If the pituitary–adrenal axis plays such an important role in central neurotransmission, it may be speculated that glucocorticoid synthesis inhibitors (for example metapyrone) could reduce the abnormality in neurotransmitter function by decreasing the cortisol concentration.

CHANGES IN NEUROTRANSMITTERS IN DEPRESSED PATIENTS FOLLOWING ANTIDEPRESSANT TREATMENT

There is convincing evidence from animal studies that, due to the anatomical and functional interconnections between the main biogenic amine neurotransmitters, the concept of specificity of an antidepressant for a particular neurotransmitter system is a gross simplification. Until recently however, it has been difficult to test the validity of such findings in depressed patients. The advent of position emission tomographic methods may enable the hypothesis to be tested directly but, to date, reliance has been placed on data obtained from an analysis of amine metabolites in the CSF of depressed patients before and following effective antidepressant treatment.

The selectivity of selective antidepressants in depressed patients was first challenged by Montgomery (1982). In this study, a group of depressed patients that had either normal to reduced CSF 5-hydroxyindole acetic acid concentrations were randomly allocated to treatments with either the specific 5-HT uptake inhibitor zimelidine or the noradrenalaine uptake inhibitor maprotiline. The results of this study showed that the response to antidepressant treatment was unconnected with the 5-HT status of the patient.

Potter *et al.* (1989) have summarized the results of seven studies by different investigators in which the effects of different tricyclic (imipramine, desipramine, amitriptyline, nortriptyline, and clomipramine) and non-tricyclic (zimelidine and citalopram) antidepressants were studied for their effects on the CSF concentrations of the main 5-HT metabolite (5-hydroxyindole acetic acid), the noradrenaline metabolite (methoxy hydropxyphenylglycol), and the dopamine metabolite (homovanillic acid) in depressed patients. The results of these studies showed that irrespective of the acute specificity of these antidepressants for the noradrenergic (for example desipramine, nortriptyline) or serotonergic (for example clomipramine, zimelidine, citalopram) systems, all produced a qualitatively similar decrease in the 5-HIAA and MHPG concentrations. The slight rise in CSF HVA concentrations occurred after the administration of those drugs that showed some selectivity for the serotonergic system. Hsiao *et al.* (1987) also investigated the inter-relationship between the same mono-amine metabolites in the CSF of depressed patients who either responded or who failed to respond to antidepressant treatment. They found that those responding to treatment showed correlations between the CSF meta-bolites that were similar to those found in control subjects. Conversely, those patients failing to respond to treatment did not show such correlations.

Thus the results of clinical studies appear to confirm the conclusions reached from observations based on the effects of chronic antidepressant treatments in animals. The changes in central serotonergic function induced by all antidepressants would appear to be a necessary though not a sufficient action for therapeutic activity. In this regard, it is of interest that lithium augmentation of antidepressants has been used to treat those depressed patients who fail to respond to conventional antidepressant treatment (de Montigny *et al.* 1985). Price *et al.* (1986) have shown that the rapid therapeutic response (< 7 days) to lithium augmentation is less common than originally suggested. This suggests that the ability of lithium to facilitate 5-HT release may not be the reason for its therapeutic action but that the subsequent changes in 5-HT receptors, and possibly dopaminergic function, more closely correlates with the onset of the therapeutic response. Thus lithium, like the antidepressants, may act by stabilizing homeostatic mechanism that are dysregulated in patients suffering from affective disorders (Rudorfer *et al.* 1985).

Perhaps the time is opportune to move from the nosological to the functional approach when undertaking research into the affective dis-orders. In this respect, van Praag and co-workers (1990) have convincingly argued that signs of abnormal dopaminergic, serotonergic, and nora-drenergic function occur in various psychiatric disorders. These changes are not disorder specific but rather they are related to the psychopatho-logical state (for example hypoactivity and inertia, increased aggressive-ness and anxiety, anhedonia) and occur independently of the nosological framework in which such neurotransmitter dysfunctions occur.

ENDOGENOUS ENDOCOIDS, DEPRESSION AND THE THERAPEUTIC EFFECT OF ANTIDEPRESSANTS

Is it possible that the changes in serotonergic function in depressed patients are attributable to the presence of endogenous factor(s) that initiate the pathological changes? Such a concept is quite plausible when one considers the role of endogenous opioids in affecting the mood and pain perception, or the possible role of endogenous ligands that act on the benzodiazpeine receptors to alleviate or cause anxiety.

Brusov and co-workers (1985) showed that the plasma and brain contain factors that modulate the action of 5-HT on platelet membranes. These investigators showed that in a population of drug-free unipolar and bipolar depressed women, the concentration of 5-HT sufficient to induce half maximal platelet shape change (a measure of $5-HT_2$ receptor function) was significantly lower than in a group of age-and sex-matched controls. Healy *et al.* (1983, 1985) have also shown that effective antidepressant treatment results in a normalization in platelet $5-HT_2$ receptor function. The results of such studies suggest that abnormal $5-HT_2$ receptor function in the depressed patient may be due to the presence of endogenous peptides(s) (Roth *et al.* 1985). Such peptides have been detected in bovine forebrain and inhibit the binding of both 3H-mianserin and 3H-ketanserin to $5-HT_2$ receptors (Roth *et al.* 1985).

Other endogenous factors have recently been identified in bovine brain and human platelet extracts that inhibited 3H-5-HT uptake and 3H-imipramine binding to human platelets (Tang *et al.* 1990). Some of these low molecular-weight endogenous substances were recognized by rabbit antibodies raised against imipramine. Thus it would appear that endogenous modulation of 5-HT uptake exist in brain and platelets that possess a partial molecular structure which is similar to that identified by the imipramine antibodies. Whether such substances are identical to those reported to interfere with $5-HT_2$ receptor function, or the binding of mianserin and imipramine to their binding sites as has been reported by other investigators, is unclear.

Studies in our laboratory have recently shown that the aggregatory response of platelets to 5-HT was reduced by a plasma factor that only occurred in depressed patients during the active phase of the illness. This plasma factor(s) was also found to inhibit the aggregatory response induced by 5-HT in control subjects (Leonard 1993). Once the patients recovered from depression, the plasma factor no longer affected the 5-HT induced aggregation. The plasma from depressed patients was also shown to reduce the rate of 3H-5-HT uptake into platelets from control subjects. While the identity of this plasma factor(s) is unknown, possible candidates include various glycoproteins and prostenoids. Thus alpha acid-1 glycoprotein has been shown to be elevated in depressed patients and normalizes following effective antidepressant treatment (Nemeroff *et al.* 1990). It should be noted however that not all investigators have been able to replicate such findings (Healy *et al.* 1991) so that the precise identity of the plasma factor remains an enigma.

It may be speculated that antidepressants produce their therapeutic effects not primarily by specifically modulating the activity of dysfunctioning neurotransmitter systems but rather by counter-acting the adverse effects of endogenous endocoids. Could this factor be the 'black bile' which has preoccupied clinicians since the time of Hippocrates?

CONCLUSION

It appears that the chemically and pharmacologically diverse group of drugs that act as antidepressants have two things in common. Firstly, they are of approximately equal clinical efficacy and take several weeks to produce an optimal therapeutic effect. Secondly, they all modulate a number of different types of mainly postsynaptic neurotransmitter receptors in animals and in depressed patients. Such effects occur irrespective of their specific effects on a particular neurotransmitter system that may arise following their acute administration. Thus the amine hypothesis of depression, in which it was postulated that depression arises from a deficiency in one or more of the biogenic amine neuotransmitters which is corrected following effective antidepressant treatment, has now evolved into the receptor sensitivity hypothesis. This hypothesis explains the delay in the therapeutic effects of antidepressant treatments as being due to time-dependent adaptational changes in neurotransmitter receptors. Thus depression may be related to an underlying abnormality in neurotransmitter receptor function. Whether the abnormality in receptor function is both the necessary and sufficient cause of the abnormal affect state is still a matter of debate. It is possible, for example, that the changes in receptor function that ultimately produces the behavioural changes associated with depression are a consequence of changes in central glucocorticoid receptor activity that reflect the response of the patient to prolonged hypercortisolaemia.

Clearly more research is needed to elucidate the precise mechanism of the action of antidepressants and, more particularly, obtain a fuller understanding of the biology of depression. Hopefully, the wider use of various imaging techniques, quantitative receptor autoradiography, etc. in patients will greatly assist in this task. In experimental studies, more specific probes need to be developed not only for neurotransmitter receptors but also for secondary and tertiary messengers. Finally, the possible role of such neuropeptides as somatostatin, neuropeptide Y, and corticotorphic releasing factor, which are closely associated with the biogenic amine neurotransmitters, need particular consideration.

REFERENCES

Aghajanian, G.K. and de Montigny, C. (1978). Tricyclic antidepressants: long term treatment increases responsivity of rat forebrain neurons to serotonin. *Science*, **202**, 1303–5.

Agren, H., Mefford, I.N., Rudorfer, M.V., Linnoila, M., and Potter, ZW.Z. (1986). Interacting neurotransmitter systems. *Journal of Psychiatric Research*, **20**, 175–93.

Ayd, F.J. and Blackwell, B. (ed.) (1970). *Discoveries in biological psychiatry*. Lipincott, Philadelphia.

Berridge, M.J. (1987). Inositol triphosphates and diacyl glycol: two interacting second messengers. *Annual Review of Biochemistry*, **56**, 159–93.

Blier, P., Chaput, Y., and de Montigny, C. (1988). Long-term 5-HT reuptake blockade, but not monoamine oxidase inhibition, decreases the function of the terminal 5-HT autoreceptors: an electrophysiological study in the rat brain. *Naunyn-Schmiedebergs Archives of Pharmacology*, **337**, 246–54.

Brunello, N., Volterra, A., Cagiano, R., Ianiere, G.C., Cuoma, V., and Racagni, G. (1985). Biochemical and behavioral changes in rats after prolonged treatment with desipramine: interaction with p-chlorophenylalanine. *Naunyn-Schmiedebergs Archives of Pharmacology*, **331**, 20–2.

Brusov, O.S., Beliaev, B.S., Katasonov, A.B., Zlobina, G.P., Factor, M.I., and Lideman, R.R. (1989). Does platelet serotonin receptor supersensitivity accompany endogenous depression? *Biological Psychiatry*, **25**, 375–81.

Burnstein, K.L. and Cidlowski, J.A. (1989). Regulation of gene expression by glucocorticoids. *Annual Review of Physiology*, **51**, 683–99.

Butler, J. and Leonard, B.E. (1990). Comparison of mianserin and dothiepin treatment on the serotonergic system in depressed patients. *Human Psychopharmacology*, **5**, 369–72.

Chaput, Y. and de Montigny, C. (1988). Effects of the 5-hydroxytryptamine antagonist BMY7378 on 5-hydroxytryptamine neurotransmission: electrophysiological studies in the rat central nervous system. Journal of Pharmacology and Experimental Therapeutics, **246**, 359–70.

Charney, D.S., Southwick, S.M., Delgado, P.L., and Krystal, H.H. (1990). Current status of the receptor sensitivity hypothesis of antidepressant action. In *Pharmacotherpay of depression* (ed. J.D. Amsterdam), pp. 13–34. Marcel Dekker, New York.

de Montigny, C., Elie, R., and Caille, G. (1985). Rapid response to the addition of lithium in iprindole-resistant unipolar depression: a pilot study. *American Journal of Psychiatry*, **142**, 220–3.

Dray, A., Gonge, T.J., and Oakley, N.R. (1976). Evidence for the existence of a raphe projection to the substantia ngira in rat. *Brain Research*, **113**, 45–57.

Earley, B., Glennon, M., Lally, M., Leonard, B.E., Junien, J.L., and Westein, J. (1990). Autoradiographic distribution of cholinergic muscarinic receptors in olfactory bulbectomized rats after chronic treatment with mianserin and desipramine. *Proceedings of the Society of Neuroscience, Abstract* No 1.

Everett, G.M. and Tolman, J.E.O. (1959). Mode of action of Rauwolfia allcalloids and motor activity. In *Biological psychiatry* (ed. J. Masserman), pp. 75–87. Grune and Stratton, New York.

Giller, E.L., Yehuda, R., and O'Loughlin, M. (1991). Lymphocyte glucocorticoid receptors in PTSD, major depression, panic and schizophrenic. *Proceedings of the American College of Neuropsychopharmacology Meeting*, San Juan, p. 179.

Harfstrand, A., Fuxe, K., and Cintra, A. (1986). Glucocorticoiod receptor immuno reactivity in monoamine neurons of rat brain. *Proceedings of the National Academy of Sciences USA*, **83**, 9779–83.

Heal, D.J. (1990). The effects of drugs on behavioural models of central noradrenergic function. In *The pharmacology of noradrenaline in the central nervous system* (ed. D. Heal and C.A. Marsden), pp. 266–315. Oxford Medical.

Healy, D., Carney, P.A., and Leonard, B.E. (1983). Monoamine related markers of depression. *Journal of Psychiatric Research*, **17**, 251–8.

Healy, D., Carney, P.A., O'Halloran, A., and Leonard, B.E. (1985). Peripheral adrenoceptors and serotonin receptors in depression. *Journal of Affective Disorders* **17**, 285–92.

Healy, D., Calvin, J., Whitehouse, A.M., White, W., Wilton-Cox, H., Theodorou, A.E. *et al.* (1991). Alpha-1-acid glycoprotein in major depressive and eating disorders. *Journal of Affective Disorders*, **22**, 13–20.

Hsiao, J.K., Agren, H., Bartko, J.J., Rudorfer, M.V., Linnoila, M., and Potter, W.Z. (1987). Monoamine neurotransmitter interactions and the prediction of antidepressant response. *Archives of General Psychiatry*, **44**, 1078–83.

Kitayama, I., Janson, A.M., and Cintra, A. (1988). Effects of chronic imipramine treatment on glucocorticoid receptor immuno reactivity in various regions of the rat brain. *Journal of Neural Transmission*, **73**, 191–203.

Leonard, B.E. (1993). In search of black bile: do antidepressants act by changing endogenous endocoids in the depressed patient? *Journal of Psychopharmacology*, **7**, 1–3.

Montgomery, S.A. (1982). The non-selective effects of selective antidepressants. *Advances in Biochemistry and Psychopharmacology*, **32**, 49–58.

Nemeroff, C.B., Krishnan, R.R., Blazer, D.G., Knight, D.L., Benjamin, D., and Meyerson, L. (1990). Elevated plasma concentration of alpha 1-1-acid glycoprotein, a putative endogenous inhibitor of the tritiated imipramine binding site, in depressed patients. *Archives of General Psychiatry*, **47**, 337–40.

Nestler, E.J., Walaas, S.T., and Greengard, P. (1984). Neuronal phosphoproteins: physiological and clinical implications. *Science*, **225**, 1357–64.

Oswald, I., Brezinova, V., and Dunleavy, D.L. (1972). On the slowness of action of tricyclic antidepressant drugs. *British Journal of Psychiatry*, **120**, 673–7.

Potter, W.Z., Hsiao, J.K., and Agren, H. (1989). Neurotransmitter interactions as a target of drug action. In *Clinical pharmacology and psychiatry* (ed. S.G. Dahl and L.F. Gram), pp. 40–51. Springer, Berlin.

Price, L.H., Charney, D.S., and Heninger, G.R. (1986). Variability of response to lithium augmentation in refractory depression. *American Journal of Psychiatry*, **143**, 1387–92.

Racagni, G., Tinelli, D., Bianchi, E., Brunello, N., and Perez, J. (1991). cAMp-dependent binding proteins and endogenous phosphorylation after antidepressant treatment: In *5-Hydroxytryptamine in psychiatry* (ed. M. Sandler, A. Coppen, and S. Harenet), pp. 116–23. Oxford Medical.

Roth, B.L., Nakaki, T., Chauang, D.M., and Costa, E. (1985). Evidence for an endocoid for the 5HT2 recognition site. *Progress in Clinical and Biological Research*, **192**, 473–6.

Rudorfer, M.V., Karoum, F., Ross, R.J., Potter, W.Z., and Linnoila, M. (1985). Differences in lithium effects in depressed and healthy subjects. *Clinical and Pharmacological Therapy*, **37**, 66–71.

Serra, G., Argiolas, A., Fadda, F., and Gessa, G.L. (1980). Hyposensitivity of dopamine autoreceptors induced by chronic administration of tricyclic antidepressants. *Pharmacological Research Communicators*, **12**, 619–25.

Stockmeier, C.A. and Keller, K.J. (1986). *In vivo* regulation of the serotonin 2 receptor in rat brain. *Life Science*, **38**, 117–27.

Sulsler, F. and Sanders-Bush, E. (1987). In *Molecular basis of neuronal responsiveness* (ed. Y.H. Ehrlich, R.H. Lenox, E. Kornecki, and W.O. Berry), pp. 489–502. Plenum Press, New York.

Tang, S.W., Cheung, S., Strijewski, A., Chudzik, J., and Helmeste, D. (1990). Anti-imipramine antibodies recognize endogenous serotonin uptake and imipramine binding inhibitors. *Psychiatry Research*, **34**, 205–12.

van Praag, H.M., Asnis, G.M., Kahn, R.S., Brown, S.L., Korn, M., Harkavy Friedman, J.M. *et al.* (1990). Monoamines and abnormal behaviour. *British Journal of Psychiatry*, **157**, 723–34.

Wong, K.L., Bruck, R.C., and Farbman, A.I. (1991). Amitriptyline mediated inhibition of neurite outgrowth from chick embryonic cerebal explants involves a reduction in adenylate cyclase activity. *Journal of Neurochemistry*, **57**, 1223–30.

3

Animal studies on neuropeptides and depression

JAN M. VAN REE, CÉCILE DURLACH-MISTELI, and
RAYMOND J.M. NIESINK

INTRODUCTION

Pituitary hormones regulate the endocrine organs and are involved in many homeostatic mechanisms in the body. They have also direct effects on processes in the central nervous system. This was demonstrated using a variety of techniques, including behavioural, neurochemical, and electro-physiological ones (van Ree et al. 1978; De Wied and Jolles 1982). Most of these effects on brain functions can be mimicked by small segments of the molecules, which are devoid of the classical endocrine activities of the parent hormone (van Ree et al. 1978; De Wied and Jolles 1982). Peptide molecules affecting the nervous system or present in neural tissue have been designated as neuropeptides (De Wied et al. 1974). Research during the last decades has disclosed that many neuropeptides are present in the central nervous system, and that they are presumably located in neuronal pathways. They are synthesized in large proteins. Enzymatic processing of these precursor molecules results in the generation of biologically active peptides that may influence brain function. These peptides in turn are susceptible to enzymatic activity, yielding other neuropeptides. The changing bioavailability of neuropeptides, probably controlled by enzymatic processing, allows the peptidergic systems to play an unique role in the regulation of brain homeostatic mechanisms.

The enzymatic processes may control the quantities of neuropeptides synthesized as well as the nature of their biological activity through size, form, and derivation of the end product. In this way sets of neuropeptides with different, opposite, and more selective properties are formed from the same precursor.

The symptomatology of depression or affective disorders is rather hetero-genous. The main symptoms include mood changes and disturbances in appetite, energy, libido, sleep, and circadian rhythms. The latter functions are at least partly controlled by brain systems located in the hypothalamus and lower brainstem. Many of these centres are known to be innervated by particular peptidergic systems. In addition, neuroendocrine dysfunctions have been demonstrated in depressed patients, for example diminished suppression of cortisol after dexamethason treatment (Carroll 1985) and a blunted thyroid-stimulating hormone (TSH) response to thyrotropin-releasing hormone (TRH) (Loosen and Prange 1982). The mood disturbances in depressed patients can be accompanied by cognitive dysfunctions (Johnson and Magaro 1987). It has been proposed that peptides related to adrenocorticotropic hormone (ACTH) (De Wied and Jolles 1982) and vasopressin (van Ree *et al.* 1978) are implicated in cognitive processes, and that the opioid peptide β-endorphin, which like ACTH is derived from the precursor molecule pro-opiomelanocortin, may be involved in mood changes (van Ree 1982). Thus, there may be a relation between depression and neuropeptides, in particular those related to pituitary hormones and the hypothalamic peptides controlling the release of these hormones.

STRATEGIES IN ANIMAL RESEARCH ON DEPRESSION

Human subjects and experimental animals differ markedly as to living environment and behaviour. However, basic principles implicated in behavioural regulation may differ less. This could permit certain effects of behaviourally active peptides in animals to be reproduced in humans. Since neuropeptides are implicated in the control of a variety of brain functions, they may also be involved in disturbances of these functions and consequently be used for counteraction of these disturbances. To make predictions for brain diseases, three strategies can be followed in animal experiments, i.e. (patho)physiological, pharmacological, and pathological approaches. These strategies may also contribute to the unravelling of possible relations between endogenous substances, including neuropeptides, and depression.

The physiological approach includes the investigation of physiological processes in the brain, that are disturbed in depression. These processes may be related to one or a set of symptoms characteristic for the disease. Knowledge about substances and mechanisms implicated in the regulation of these processes may eventually lead to entities with antidepressant action. This approach is, however, limited due to the heterogenous symptomatology of the depressive illness and because a key symptom can hardly be designated. One characteristic feature is mood changes. Although it is rather difficult to assess mood in animals, it has been suggested that intracranial self-stimulation, investigating the brain systems that mediate reward, may bear some relation with mood in humans. This behavioural procedure also fulfils other validating criteria for an animal model for depression (Willner 1984). Symptoms frequently accompanying depression are disturbances of cognitive functions. Thus, animal studies aimed at improvement of cognition may be relevant for the development of drugs for depression.

The pharmacological approach involves, first the characterization of the actions in animals of drugs known to be effective in treating depressive patients, and second, the comparison of the effects of new entities with those of the antidepressants. Most of the research so far has been concentrated on this approach. But little is known about the relation between the effects of antidepressants in animals and the therapeutic action of these drugs in patients. Moreover, most of the presently available models deal with acute effects of antidepressants, whereas their therapeutic action is visible after chronic treatment only. And not all effective antidepressant treatment, including the second generation of antidepressants and electroconvulsive shock, shares the same action in the models (Willner 1984). Notwithstanding these remarks, the pharmacological approach is still widely used to develop new antidepressants, but until now a new class of other drugs has not been generated by this approach.

The third approach deals with pathophysiology and pathology. It includes the induction of the similar pathology in animals as is present in patients. Information about the pathological processes in depressed patients is, however, barely available and this hampers the development of models in this respect. Most animal studies dealing with this approach have concentrated on phenomena related to stress. Examples are maternal separation, reversal of the light–dark cycle, exposure to chronic unpredictable stress, learned helplessness, and behavioural despair. In most of these models an activation phase is followed by an inhibition phase. Interestingly, this last phase may be associated with a decrease in positively reinforced behaviour, which characterizes the intracranial self-stimulation model (Willner 1984).

A number of studies with neuropeptides following the three different approaches are described in the next sections.

NEUROPEPTIDES AND THE PHYSIOLOGICAL APPROACH

Characteristic symptoms of depression are, among others, mood changes and a decrease of the ability to experience pleasure. An animal model dealing with so called pleasure centres is intracranial electrical self-stimulation (ICSS), investigating brain systems which mediate reward (Wauquier and Rolls 1976). Brain stimulation reward has been connected with mood changes and especially mood elevation (Routtenberg 1978; Clavier and Routtenberg 1980). Furthermore, commonalities between reward from brain stimulation, dependence-inducing drugs, and natural reinforcers have been suggested (Wise 1980). A number of theories of depression focuses on the postulate that depression results from a reduction in the activity of reward systems (Costello 1972; Lewinsohn *et al.* 1979). Thus, the analysis of brain stimulation reward can contribute to animal research on depression, bearing as a model some predictive, face, and construct validity (Willner 1984). Of interest is the report showing that chronic administration of desipramine increased the sensitivity of rats to brain stimulation reward (Fibiger and Philips 1981).

One of the brain systems implicated in the brain reward circuitry is the dopaminergic pathway with cell bodies in the ventral tegmental area (VTA) and terminals in the nucleus accumbens among others (Wise 1983). Some peptide studies have been performed with ICSS in animals implanted with electrodes in the VTA. The role of endogenous opioids has been investigated using the opioid antagonist naloxone. Acute treatment with this drug increased the threshold for self-stimulation, using response-rate sensitive and insensitive procedures (van Wolfswinkel and van Ree 1985). This effect was more pronounced after repeated treatment (van Wolfswinkel et al. 1985). Of particular interest is that the increase in threshold persisted after discontinuation of naloxone treatment, suggesting that endogenous opioids are involved in the setpoint of brain stimulation reward (van Wolfswinkel et al. 1985). An increased threshold for self-stimulation may result in difficulties in enjoying normally pleasant situations, assuming the existence of a relation between brain stimulation reward and natural reinforcements. Further, it might be postulated that the setpoint of brain stimulation reward is important for the setpoint of mood. Thus, a decreased activity of endogenous opioids may result in anhedonia and mood changes.

Peptides related to neurohypophyseal hormones modulate self-stimulation of the VTA. Thus, the vasopressin fragment DGAVP diminishes, while the C-terminal fragment of oxytocin, prolyl-leucylglycinamide, increases the response rate (Dorsa and van Ree 1979). A similar modulation has been reported for fragments of β-endorphin. This opioid peptide is converted to α- and γ-type endorphins by brain enzymes (Burbach et al. 1980). α-Type endorphins induce effects that are also observed with the psychostimulant drug amphetamine, and the effects of γ-type endorphins bear similarities with those of neuroleptic drugs like haloperidol (De Wied et al. 1978; van Ree et al. 1980). Such comparative actions have also been observed with self-stimulation of the VTA (Dorsa et al. 1979). Thus, α-endorphin and amphetamine increase, while des-Tyr[1]-γ-endorphin and haloperidol diminish self-stimulating behaviour. Further studies, for example chronic treatment with these peptides, are needed before definite conclusions can be drawn concerning the relevance of the mentioned effects of self-stimulation for depression. But, the data show that self-stimulation of the VTA is sensitive to the modulatory action of neuropeptides from different sources.

Cognitive dysfunctions may accompany the mood disturbances in depression. Thus, entities with a cognition-enhancing action may at least be useful in treating these dysfunctions. Among the peptides that have been implicated in certain cognitive processes are those related to adreno-corticotropic hormone (ACTH) and vasopressin. ACTH-related peptides, including the ACTH-(4–9) analogue ORG 2766, may facilitate motivation and vigilance, enhance concentration and (visual) attention, promote learning, and stimulate memory-retrieval processes (De Wied and Jolles 1982). Vasopressin and related peptides have been implicated in memory consolidation and retrieval processes in particular (De Wied *et al.* 1988). Originally, most studies with vasopressin neuropeptides have been per-formed with aversively motivated tasks. But the memory enhancing effects could also be demonstrated in food-reinforced tasks (De Wied *et al.* 1988) and on social memory (Dantzer *et al.* 1987; Popik *et al.* 1991; Sekiguchi *et al.* 1991).

NEUROPEPTIDES AND THE PHARMACOLOGICAL APPROACH

The number of studies comparing the effects of neuropeptides with those of antidepressants are limited. One series of experiments has concentrated on dopamine and melatonin. Although most postulates concerning de-pression and antidepressants focus on noradrenaline and serotonin, a role for dopamine has been suggested as well (Willner 1981). Besides that this neurotransmitter may play a crucial role in the inability to experience pleasure, as discussed before, chronic but not acute treatment with anti-depressants may facilitate dopaminergic function in the brain (for refer-ence see Willner 1981; Durlach-Misteli and van Ree 1992). Melatonin has been implicated in the aetiology of depression and altered secretion of melatonin in depressed patients has been reported (Levy *et al.* 1979; Mendlewicz *et al.* 1979; Wetterberg *et al.* 1979; Braun *et al.* 1985). Animal studies have suggested that melatonin plays an important role in certain behavioural processes, especially in modulation of locomotor activity and emotionality (Kovacs *et al.* 1974; Datta and King 1980*a*; Golus and King 1981). These behavioural patterns may be mediated by the mesolimbic dopaminergic projections to the nucleus accumbens (Costall *et al.* 1977; Blanc *et al.* 1981; van Ree and Wolterink 1981). Acute and chronic treat-ment with anti-depressants resulted in enhanced levels of melatonin in the pineal gland and plasma (Wirtz-Justice *et al.* 1980; Heydorn *et al.* 1982). Moreover, an interaction between melatonin and dopamine has been sug-gested (Datta and King 1980*b*; Zisapel *et al.* 1982). Therefore, some studies were performed to investigate the link between antidepressants, melatonin, and the mesolimbic dopamine system.

Small doses of melatonin injected into the nucleus accumbens of rats decreased locomotor activity and rearing, and increased grooming and sniffing behaviour (Gaffori and van Ree 1985a). Larger doses of melatonin appeared to be less effective. The melatonin-induced behavioural responses were completely inhibited by local treatment with small doses of various antidepressant drugs, i.e. zimelidine, mianserin, nortriptyline, clomipramine, and desipramine. This effect of the antidepressants was rather specific, since they did not interfere with the decrease of locomotor activity and rearing induced by injection of small doses of the dopamine agonist apomorphine into the nucleus accumbens. In contrast, dopamine antagonists like haloperidol and sulpiride did not affect the melatonin-induced behavioural changes, but antagonized the decreased motor activity induced by small doses of apomorphine (van Ree *et al.* 1982; Gaffori and van Ree 1985a). The effect of melatonin injected into the nucleus accumbens was mimicked by serotonin antagonists (methysergide and cyproheptadine) and inhibited by serotonin, suggesting an inter-relationship between melatonin and serotonin systems in the nucleus accumbens. Since terminals of the pro-opiomelanocortin systems are present in the nucleus accumbens (Khatchaturian *et al.* 1985) and γ-type endorphins, fragments of β-endorphin, inhibit the apomorphine-induced hypolocomotion (van Ree *et al.* 1982), β-endorphin and several of its fragments were tested for their possible interaction with melatonin-induced behavioural changes. It was found that both α- and γ-endorphins, but not β-endorphin antagonized the melatonin-induced responses, following injection into the nucleus accumbens (Gaffori and van Ree 1985b). Structure–activity studies revealed that the active moiety resides in the fragment β-endorphin (10–16) (βE-(10–16)). This peptide did not affect the apomorphine-induced hypolocomotion. Thus, the profile of this peptide with respect to melatonin- and apomorphine-induced responses after acute treatment into the nucleus accumbens, resembles that of antidepressants.

Next, the effect of chronic systemic treatment with antidepressants and βE-(10–16) was studied on the behavioural effects of melatonin and apomorphine injected into the nucleus accumbens (Durlach-Misteli and van Ree 1992, and unpublished observations). As antidepressants were selected desipramine, a preferentially noradrenaline-reuptake inhibitor, fluvoxamine, a preferentially serotonin-reuptake inhibitor, and mianserin, a tetracyclic antidepressant with mixed action on various amine systems (Durlach-Misteli and van Ree 1992). The rats were first equipped with a permanent cannula on each side of the brain and aimed at the nucleus accumbens. After about one week the animals received a chronic systemic treatment, once daily for about 20 days (desipramine 15 mg/kg, i.p.; fluvoxamine 15 mg/kg, s.c.; mianserin 15 mg/kg, i.p; βE-(10–16), 50 micrograms per animal (about 0.25 mg/kg, s.c.). All experiments included a group of animals treated with placebo. After 2 weeks of treatment (days 13–16 after starting treatment) the animals were injected with saline or graded doses of apomorphine and subsequently the behaviours of the rats were observed. The doses of apomorphine ranged from 1 ng to 10 micrograms (five doses) to test the sensitivity of both pre- and postsynaptic dopaminergic receptor systems (van Ree and Wolterink 1981; van Ree *et al.* 1982, 1983). The rats were injected into the nucleus accumbens with saline and melatonin (10 or 100 ng) on day 18 or 19 after starting treatment.

The sensitivity of presynaptically located dopaminergic receptor systems, as evidenced by the behavioural response upon challenge with low doses of apomorphine (i.e. decreased locomotion and rearing), was differentially altered by the three antidepressants tested. This sensitivity was not changed by chronic pretreatment with desipramine, enhanced by fluvoxamine, and attenuated by mianserin. However, treatment with the three antidepressants resulted in an increased hypermotility response upon challenge with the high doses of apomorphine. These data suggest that the chronic treatment with antidepressants is accompanied by development of supersensitivity of postsynaptic dopaminergic receptor systems. The melatonin-induced behavioural changes, i.e. decreased motility and rearing, and increased sniffing, were completely absent in animals chronically treated with desipramine, fluvoxamine, or mianserin, as was observed after acute treatment with antidepressants (Gaffori and van Ree 1985*a*). Thus, antidepressants block the action of melatonin in the nucleus accumbens both after acute and chronic treatment. It is tempting to speculate that the antimelatonin action mediates the dopaminergic sensitization process, which may be responsible for the delayed onset of the antidepressant action of antidepressant drugs. Whether the interaction between the antidepressants and melatonin is mediated by serotonin or not, is yet unclear.

The peptide βE-(10–16) mimicked the action of antidepressants in the aforementioned test procedures after chronic treatment. Low doses of apomorphine decreased motility (Fig. 3.1) and rearing (not shown), while higher doses increased both parameters. The effect of low doses was somewhat enhanced in animals pretreated with βE-(10–16), and the hypermotility response after a high dose of apomorphine was potentiated. Melatonin injections resulted in decreased motility and increased sniffing behaviour (Fig. 3.2). These responses were not present in rats treated chronically with the peptide, as was observed after acute treatment with βE-(10–16) (Gaffori and van Rec 1985*b*). Thus, according to the pharmacological isomorphism principle, the peptide βE-(10–16) may be a potential antidepressant-like peptide, although more studies are needed to substantiate this postulate.

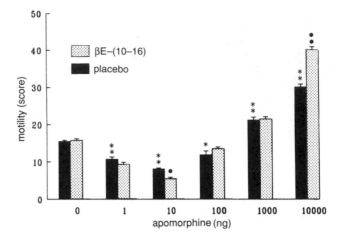

FIG. 3.1. Chronic systemic treatment with the β-endorphin fragment βE-(10–16) and the responsiveness to the dopamine agonist apomorphine injected into the nucleus accumbens. Groups of rats ($n = 6$–14) were pretreated with βE-(10–16) 50 micrograms s.c. once daily for 14 days or placebo and bilaterally injected with saline (0) or graded doses of apomorphine (1–10.000 ng). Rats were tested for 3 min in a small open field 20 min after injection and their behaviours were scored. Depicted are the motility scores versus the dose of apomorphine. Rearing followed the same pattern as motility, while sniffing behaviour was not changed by apomorphine treatment. Values are expressed as means ± SEM. Statistical analysis was performed by two-way ANOVA followed by Student *t*-tests. (* difference between saline and apomorphine treatment in rats pretreated with placebo (*$p <$ 0.01, **$p < 0.001$). ★ difference between placebo and βE-(10–16) pretreatment (★$p < 0.01$, ★★$p < 0.001$).

NEUROPEPTIDES AND THE PATHO(PHYSIO)LOGICAL APPROACH

The pathological approach has hardly been applied in studies on neuro-peptides and depression. Some studies have been performed that bear elements of this approach but also of the physiological and pharmacological approaches. An example is the effect of short-term isolation on social behaviours of rats. An isolation period of 7 days results in increased social behaviours (social exploration behaviours, contact behaviours (crawl over, mount, social grooming) and approach–follow), when rats are tested in a dyadic encounter (Niesink and van Ree 1982, 1984). A single treatment with the antidepressant drugs clomipramine, nortriptyline, and mianserin normalized the increased social interactions of the isolated rats to the level of the group-housed rats. The social behaviours of group-housed rats were

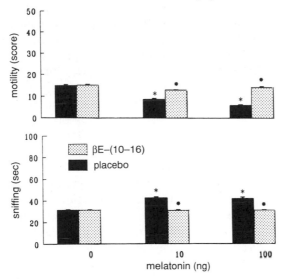

FIG. 3.2. Chronic systemic treatment with the β- endorphin fragment βE-(10–16) and the responsiveness to melatonin injected into the nucleus accumbens. Groups of rats (*n* = 8) were pretreated with βE-(10–16) 50 micrograms s.c. once daily for 19 days or placebo and bilaterally injected with saline (0) or melatonin (10 or 100 ng). Rats were tested for 3 min in a small open field 45 min after injection and their behaviours were scored. Depicted are the motility scores and the duration of sniffing behaviour (s) versus the dose of melatonin. Rearing followed the same pattern as motility. Values are expressed as mean ± SEM. Statistical analyses were performed by two-way ANOVA followed by Student *t*-tests. * difference between saline and melatonin treatment in rats pretreated with placebo (*p* < 0.001). ★ difference between placebo and βE-(10–16) pretreatment (*p* < 0.001).

not affected (Niesink and van Ree 1982). An opiate (morphine), an opiate antagonist (naloxone), a tranquillizer (diazepam), a neuroleptic (halo-peridol), and a psychostimulant (amphetamine) did not preferentially in-fluence the social behaviours of the isolated rat. Chronic treatment with the antidepressants did not reduce the increased social interactions of the isolated animals. In spite of this, the short-term social isolation procedure might be useful for predicting antidepressant activity. A number of neuro-peptides were tested in this model. It was found that the ACTH-(4–9) analogue ORG 2766, the C-terminal tripeptide of oxytocin (PLG) and thyrotropin-releasing hormone (TRH) counteracted the increase in social behaviours due to short term isolation (Niesink and van Ree 1984). Other peptides were not effective in this respect, but most of these peptides were administered using one dose only. These data suggest that the mentioned effective peptides—ORG 2766, PLG, and TRH—mimic the action of antidepressants in this model. Whether the brain mechanisms underlying the effects of short-term isolation are also involved in other models of isolation (maternal separation (Everitt and Keverne 1979), hyperactivity after long-term isolation (Garzon and Del Rio 1981)) remains to be studied.

Circadian oscillations have been reported for a number of brain functions that are more or less implicated in memory processes (Barnes et al. 1977). This may be pertinent for depression, in which disturbances of both circadian rhythms and cognitive functions have been suggested (Goodwin et al. 1982; Johnson and Magaro 1987). Moreover, readjustment to a normal light–dark cycle was facilitated by antidepressant treatment (Baltzer and Weiskrantz 1973) and sleep deprivation has been proposed as an antidepressant treatment in patients (Wu and Bunney 1990). Disrupting circadian organization in rats by phase-shifting the illumination cycle by exposure to a reversed light–dark cycle, resulted in retrograde amnesia for passive and active avoidance behaviour patterns (Fekete et al. 1985). This retrograde amnesia lasted 2–3 days and is probably due to retrieval disturbances. A single treatment with either the ACTH-(4–9) analogue ORG 2766 or desglycinamide-(Arg^8) vasopressin (DGAVP) prior to the retention of passive or extinction of active avoidance behaviour restored the behavioural impairment (Fekete et al. 1986). Thus, these peptides may relieve memory deficits induced by disturbances in circadian organization. The significance of this effect for depression remains to be shown.

The learned helplessness phenomenon–performance deficit in a learning task after exposure to uncontrollable stress—is one of the models frequently used to detect antidepressant activity in animals (Willner 1984). Endogenous vasopressin may be implicated in the performance deficit, since this was not observed in animals treated intraventricularly with anti-vasopressin serum (Leshner et al. 1978). Another more or less related procedure, the so called 'behavioural despair', deals with the onset of immobility when animals are forced to swim without the possibility to escape (Willner 1984). The onset of immobility in the second test is more rapid than in the first test, and is delayed by pretreatment with a wide variety of antidepressants. The action of these drugs is mimicked by the tripeptides TRH (Metcalf and Dettmar 1981; Ogawa et al. 1984) and PLG (Pulvirenti and Kastin 1988) suggesting a possible antidepressant action of the peptides. Interestingly, this effect of PLG was antagonized by the antidopaminergic drugs haloperidol and sulpiride (Pulvirenti and Kastin 1988). Another test procedure with potential construct validity for depression is chronic unpredictable stress (Willner 1984). Animals are daily subjected to different stressors during a number of weeks and subsequently their behaviours are monitored. The induced behavioural disturbances are at least partly counteracted by chronic antidepressant treatment. A similar effect has been reported after chronic treatment with the tripeptide PLG (Pignatiello *et al.* 1989).

TREATMENT OF DEPRESSED PATIENTS WITH PEPTIDES (SYNOPSIS)

The significance of hormones and peptides for depression has been evaluated using different strategies, for example the measurement of peptide levels in body fluids of patients and in post-mortem brains, peptide challenge tests and treatment of patients with peptides (see Prange and Loosen 1984; Prange *et al.* 1987). Only investigations concerning treatment of depressed patients will be mentioned briefly. Most studies have used the tripeptide TRH. This peptide was ineffective in most double-blind studies using oral administration. However, after intravenous administration, a short-lasting antidepressant effect was observed in about half of the studies, which includes more than 500 patients. The C-terminal fragment of oxytocin, PLG, has been given orally to depressed patients in four studies. In three of them, an antidepressant effect was reported which occurred rapidly and appeared to be long-lasting. An analogue of vasopressin, L-desamino-8-D-arginine-vasopressin, has been given intranasally to some depressed patients. An improved cognition was found and in some patients an antidepressant effect. A positive effect of β-endorphin has been observed, but in one out of three double-blind studies only. The opioid antagonist naloxone may worsen the symptomatology of depressed patients. The non opioid fragment of β-endorphin, des-Tyr^1-γ-endorphin (DTγE), which possesses an antipsychotic effect in a number of schizophrenic patients (van Ree *et al.* 1987), has been reported to exert an antidepressant effect in some depressed patients (Chazot *et al.* 1985).

CONCLUDING REMARKS

Research on neuropeptides using animal models of depression is limited. Still this research may be promising since a relation between depression and neuropeptides particularly those related to pituitary and hypothalamic hormones, is not unlikely. Of interest are the peptides TRH and PLG, that are active in some animal models supposing to detect antidepressant action and have been shown to exert some antidepressant activity in patients. Another peptide that mimics the effect of antidepressants after chronic administration is a fragment of β-endorphin, i.e. βE-(10–16). Investigations with this peptide and antidepressant drugs have contributed to focus on the possible relevance of the interaction between melatonin and dopamine in the nucleus accumbens for the depressive illness. Although most research so far has concentrated on noradrenaline and serotonin, evidence for a role of dopamine in depression has been provided as well (Willner 1981). The development of supersensitivity of dopaminergic receptor systems in the nucleus accumbens after chronic treatment with antidepressant drugs, could implicate subsensitivity of these systems as a psychopathological mechanism in depression. This warrants detailed research of the meso-limbic dopaminergic pathways in relation to depression and the mode of action of antidepressant drugs. Such research may include studies on intracranial self-stimulation behaviour, that have been judged as one of the reliable models of depression (Willner 1984). Some studies have shown that this behaviour is modulated by certain neuropeptides. Other peptides, for example those related to ACTH and vasopressin, affecting cognitive functions may also be useful in patients to treat at least some symptoms as observed in depression. Although not reviewed in this paper, corticotropin-releasing hormone (CRH) is also of interest. Increased levels of CRH have been observed in the cerebrospinal fluid of depressed patients (Nemeroff et al. 1984; Banki et al. 1987) and centrally administered CRH in animals caused some symptoms similar to those in the depression syndromes (Dunn and Berridge 1990). Detailed animal and human research may eventually disclose neuropeptides with therapeutic action in depression.

REFERENCES

Baltzer, V. and Weiskrantz, L. (1973). Antidepressant agents and reversal of diurnal activity cycles in the rat. *Biological Psychiatry*, **10**, 199–209.

Banki, C.M., Bissette, G., Ararto, M., O'Connor, L., and Nemeroff, C.B. (1987). CSF corticotropin-releasing factor-like immunoreactivity in depression and schizophrenia. *American Journal of Psychiatry*, **144**, 873–7.

Barnes, C.A., McNaughton, B.L., Goddard, G.V., Douglas, R.M., and Adamec, R. (1977). Circadian rhythm of synaptic excitability in rat and monkey central nervous system. *Science*, **197**, 91–2.

Blanc, G., Herve, D., Simon, H., Lipoprawski, A., Glowinski, J., and Tassin, J.P. (1980). Response to stress of mesocortico-frontal dopaminergic neurons in rats after long term isolation. *Nature*, **284**, 265–7.

Brown, R., Kocsis, J.H., Caroff, S., Amsterdam, J., Winokur, A., Stokes, P.E. *et al.* (1985). Differences in nocturnal melatonin secretion between melancolic depressed patients and control subjects. *American Journal of Psychiatry*, **142**, 811–6.

Burbach, J.P.H., Loeber, J.G., Verhoef, J., Wiegant, V.M., De Kloet, E.R., and De Wied, D. (1980). Selective conversion of β-endorphin into peptides related to γ- and α-endorphin. *Nature*, **283**, 96–7.

Carroll, B.J. (1985). Dexamethasone suppression test: a review of contemporary confusion. *Journal of Clinical Psychiatry*, **46**, 13–24.

Chazot, G., Claustrat, B., Brun, J., and Olivier, M. (1985). Rapid antidepressant activity of Des Tyr gamma endorphin: correlation with urinary melatonin. *Biological Psychiatry*, **20**, 1026–30.

Clavier, R.M. and Routtenberg, A. (1980). In search of reinforcement pathways: a neurochemical odyssey. In *Biology of reinforcement: facets of brain stimulation reward* (ed. A. Routtenberg), pp. 81–107. Elsevier, Amsterdam.

Costall, B., Taylor, R.J., Cannon, J.G., and Lee, T. (1977). Differentiation of the dopamine mechanisms mediating stereotyped behaviour and hyperactivity in the nucleus accumbens and caudate putamen. *Journal of Pharmacy and Pharmacology*, **29**, 337–42.

Costello, C.G. (1972). Depression: loss of reinforcers or loss of reinforcer effectiveness? *Behavioral Therapeutics*, **3**, 240–7.

Dantzer, R., Bluthe, R.M., Koob, G.F., and Le Moal, M. (1987). Modulation of social memory in male rats by neurohypophyseal peptides. *Psychopharmacology*, **91**, 363–8.

Datta, P.C. and King, M.G. (1980a) Melatonin: effects on brain and behavior. *Neuroscience and Behavioral Reviews*, **4**, 451–8.

Datta, P.C. and King, M.G. (1980b). Effects of MIF-1 and melatonin on novelty induced defecation and associated plasma 110HCS and brain catecholamines. *Pharmacology, Biochemistry and Behavior*, **11**, 173–81.

De Wied, D. and Jolles, J. (1982). Neuropeptides derived from pro-opiocortin: behavioral, physiological and neurochemical effects. *Physiological Reviews*, **62**, 976–1059.

De Wied, D., van Wimersma Greidanus, Tj.B., and Bohus, B. (1974). Pituitary peptides and behavior: influence on motivational, learning and memory processes. In *Neuropsychopharmacology*, pp. 653–8. Excerpta Medica, Amsterdam.

De Wied, D., Kovács, G.L., Bohus, B., van Ree, J.M., and Greven, H.M. (1978). Neuroleptic activity of the neuropeptide β-LPH$_{62-77}$ (Des-Tyr-γ-endorphin; DTγE). *European Journal of Pharmacology*, **49**, 427–36.

De Wied, D., Joëls, M., Burbach, J.P.H., De Jong, W., De Kloet, E.R., Gaffori, O.W.J. *et al.* (1988). Vasopressin effects on the central nervous system. In *Peptide hormones: effects and mechanisms of action* (ed. P.M. Conn and A. Negro Villar), pp. 97–140. CRC Press, Boca Raton.

Dorsa, D.M. and van Ree, J.M. (1979). Modulation of substantia nigra self-stimulation by neuropeptides related to neurohypophyseal hormones. *Brain Research*, **172**, 367–71.

Dorsa, D.M., van Ree, J.M., and De Wied, D. (1979) Effects of [Des-Tyr^1]-γ-endorphin and α-endorphin on substantia nigra self-stimulation. *Pharmacology, Biochemistry and Behavior*, **10**, 899–905.

Dunn, A.J. and Berridge, C.W. (1990). Physiological and behavioral responses to CRF administration: is CRF a mediator of anxiety or stress responses. *Brain Research Reviews*, **15**, 71–100.

Durlach-Misteli, C. and van Ree, J.M. (1992). Dopamine and melatonin in the nucleus accumbens may be implicated in the mode of action of antidepressant drugs. *European Journal of Pharmacology*, **217**, 15–21.

Everitt, B.J. and Keverne, E.B. (1979). Models of depression based on behavioural observations of experimental animals. In *Psychopharmacology of affective dis-*

orders (ed. E.S. Paykel and A. Coppen), pp. 41–59. Oxford University Press.

Fekete, M., van Ree, J.M., Niesink, R.J.M., and De Wied, D. (1985). Disrupting circadian rhythms in rats induces retrograde amnesia. *Physiology and Behavior*, **34**, 883–7.

Fekete, M., van Ree, J.M., and De Wied, D. (1986). The ACTH-(4–9) analog ORG 2766 and desglycinamide[9]-(Arg[8])-vasopressin reverse the retrograde amnesia induced by disrupting circadian rhythms in rats. *Peptides*, **7**, 563–8.

Fibiger, H.C. and Philips, A.G. (1981). Increased intracranial self-stimulation in rats after long-term administration of desipramine. *Science*, **214**, 683–5.

Gaffori, O. and van Ree, J.M. (1985*a*). Serotonin and antidepressant drugs antagonize melatonin-induced behavioral changes after injection into the nucleus accumbens of rats. *Neuropharmacology*, **24**, 237–44.

Gaffori, O. and van Ree, J.M. (1985*b*). β-Endorphin-(10–16) antagonizes behavioral responses elicited by melatonin following injection into the nucleus accumbens of rats. *Life Sciences*, **37**, 357–64.

Garzon, J. and Del Rio, J. (1981). Hyperactivity induced in rats by long-term isolation: further studies on a new animal model for the detection of antidepressants. *European Journal of Pharmacology*, **74**, 287–94.

Golus, P. and King, M.G. (1981). The effect of melatonin on open-field behavior. *Pharmacology, Biochemistry and Behavior*, **15**, 883–5.

Goodwin, F.K., Wirz-Justice, A., and Wehr, T.A. (1982). Evidence that pathophysiology of depression and the mechanism of action of antidepressant drugs both involve alterations in carcadian rhythms. In *Typical and atypical antidepressants: clinical practice* (ed. E. Costa and G. Racagni), pp. 1–11. Raven Press, New York.

Heydorn, W.E., Brunswick, D.J., and Frazer, A. (1982). Effect of treatment of rats with antidepressants on melatonin concentrations in the pineal gland and serum. *Journal of Pharmacology and Experimental Therapeutics*, **222**, 534–43.

Johnson, M.H. and Magaro, P.A. (1987). Effects of mood and severity on memory processes in depression and mania. *Psychological Bulletin*, **101**, 28–40.

Khachaturian, H., Lewis, M.E., Schäfer, M.K.H., and Watson, S.J. (1985). Anatomy of the CNS opioid systems. *Trends in Neurosciences*, **March**, 111–9.

Kovacs, G.L., Gajari, I., Telegdy, G., and Lissak, K. (1974). Effect of melatonin and pinealectomy on avoidance and exploratory activity in the rat. *Physiology and Behavior*, **13**, 349–55.

Leshner, A.I., Hofstein, R., Samuel, D., and van Wimersma Greidanus, Tj.B. (1978). Intraventricular injection of antivasopressin serum blocks learned helplessness in rats. *Pharmacology, Biochemistry and Behavior*, **9**, 889–92.

Lewinsohn, P.M., Yougren, M.A., and Grosscup, S.J. (1979). Reinforcement and depression. In *The psychobiology of the depressive disorders; implications for the effects of stress* (ed. R.A. Depue), pp. 291–316. Academic Press, New York.

Lewy, A.J., Wehr, I.H., Gold, P.W., and Gonduring, I.K. (1979). Plasma melatonin in maniac-depressive illness. In *Catecholamines: basic and clinical frontiers* (ed. E. Usdin), pp. 1173–85. Pergamon Press, New York.

Loosen, P.T. and Prange, A.J., Jr (1982). Serum thyrotropin response to thyrotropin-releasing hormone in psychiatric patients: a review. *American Journal of Psychiatry*, **139**, 405–16.

Mendlewicz, J., Linkowski, P., Branchey, L., Weinberg, V., Weifzman, D., and Branchey, M. (1979). Abnormal 24h pattern of melatonin secretion in depression. *Lancet*, **ii**, 1362.

Metcalf, G. and Dettmar, P.W. (1981). Is thyrotropin releasing hormone an endogenous ergotropic substance in the brain? *Lancet*, **i**, 586–9.

Nemeroff, C.B., Widerlöv, E., Bissette, G., Walleus, H., Karlsson, I., Klund, K. *et al.* (1984). Elevated concentrations of CSF corticotropin-releasing factor-like immunoreactivity in depressed patients. *Science*, **226**, 1342–3.

Niesink, R.J.M. and van Ree, J.M. (1982). Antidepressant drugs normalize the increased social behaviour of pairs of male rats induced by short term isolation. *Neuropharmacology*, **21**, 1343–8.

Niesink, R.J.M. and van Ree, J.M. (1984) Neuropeptides and social behavior of rats tested in dyadic encounters. *Neuropeptides*, **12**, 4–8.

Ogawa, N., Mizuno, S., Mori, A., Nukina, I., Ota, Z., and Yamamoto, M. (1984). Potential antidepressive effects of thyrotropin releasing hormone (TRH) and its analogues. *Peptides*, **5**, 743–6.

Pignatiello, M.F., Olson, G.A., Kastin, A.J., Ehrensing, R.H., McLean, J.H. and Olson, R.D. (1989). MIF-1 is active in a chronic stress animal model of depression. *Pharmacology, Biochemistry and Behavior*, **32**, 737–42.

Popik, P. Wolterink, G., De Brabander, H., and van Ree, J.M. (1991). Neuropeptides related to [Arg^8]vasopressin facilitates social recognition in rats. *Physiology and Behavior*, **49**, 1031–5.

Prange, A.J., Jr and Loosen, P.T. (1984). Peptides in depression. In *Frontiers in biochemical and pharmacological research in depression* (ed. E. Usdin, M. Asberg and L. Bertilsson), pp. 127–45. Raven Press, New York.

Prange, A.J., Jr, Garbutt, J.C., Loosen, P.T., Bissette, G., and Nemeroff, C.B. (1987). The role of peptides in affective disorders: a review. *Progress in Brain Research*, **72**, 235–47.

Pulvirenti, L. and Kastin, A.J. (1988). Blockade of brain dopamine receptors antagonizes the anti-immobility effect of MIF-1 and Tyr-MIF-1 in rats. *European Journal of Pharmacology*, **151**, 289–92.

Routtenberg, A. (1978). The reward system of the brain. *Scientific American*, **239**, 122–31.

Sekiguchi, R., Wolterink, G., and van Ree, J.M. (1991). Analysis of the influence of vasopressin neuropeptides on social memory of rats. *European Neuropsychopharmacology*, **1**, 123–6.

van Ree, J.M. (1982). Reward and abuse: opiates and neuropeptides. In *Brain reward systems and abuse* (ed. J. Engel, L. Oreland, D. Ingvar, B. Pernow, S. Rossner, and L.A. Pellborn), pp. 75–89. Raven Press, New York.

van Ree, J.M. and Wolterink, G. (1981). Injection of low doses of apomorphine into the nucleus accumbens of rats reduces locomotor activity. *European Journal of Pharmacology*, **72**, 107–11.

van Ree, J.M., Bohus, B., Versteeg, D.H.G., and De Wied, D. (1978). Neurohypophyseal principles and memory processes. *Biochemical Pharmacology*, **27**, 1793–800.

van Ree, J.M., Bohus, B., and De Wied, D. (1980). Similarity between behavioral effects of Des-tyrosine-γ-endorphin and haloperidol and of α-endorphin and amphetamine. In *Endogenous and exogenous opiate agonists and antagonists* (ed. E. Leong Way), pp. 459–62. Pergamon Press, New York.

van Ree, J.M., Caffé, A.R. and Wolterink, G. (1982). Non-opiate β-endorphin fragments and dopamine. III. γ-Type endorphins and various neuroleptics counteract the hypoactivity elicited by injection of apomorphine into the nucleus accumbens. *Neuropharmacology*, **21**, 1111–7.

van Ree, J.M., Gaffori, O. and De Wied, D. (1983). In rats, the behavioral profile of CCK-8 related peptides resembles that of antipsychotic agents. *European Journal of Pharmacology*, **93**, 63–78.

van Ree, J.M., Verhoeven, W.M.A., and De Wied, D. (1987). Animal and clinical research on neuropeptides and schizophrenia. *Progress in Brain Research*, **72**, 249–67.

van Wolfswinkel, L. and van Ree, J.M. (1985). Effects of morphine and naloxone on thresholds of ventral tegmental electrical self-stimulation. *Naunyn Schmiedebergs Archives of Pharmacology*, **330**, 84–92.

van Wolfswinkel, L., Seifert, W.F., and van Ree, J.M. (1985). Long-term changes in self-stimulation threshold by repeated morphine and naloxone treatment. *Life Sciences*, **37**, 169–76.

Wauquier, A. and Rolls, E.T. (1976). *Brain stimulation reward*. North Holland, Amsterdam.

Wetterberg, L., Beck-Frits, J., Aperia, B., and Petterson, V. (1979). Melatonin/cortisol ratio in depression. *Lancet*, **ii**, 1361.

Willner, P. (1981). Dopamine and depression: a review of recent evidence. *Brain Research Reviews*, **6**, 211–46.

Willner, P. (1984). The validity of animal models of depression. *Psychopharmacology*, **83**, 1–16.

Wirtz-Justice, A. Arendt, J., and Marston, A. (1980). Antidepressant drugs elevate rat pineal and plasma melatonin. *Experientia*, **36**, 442–4.

Wise, R.A. (1980). Action of drugs of abuse on brain reward systems. *Pharmacology, Biochemistry and Behavior*, **13** (Suppl. 1), 213–23.

Wise, R.A. (1983). Brain neuronal systems mediating reward processes. In *The neurobiology of opiate reward processes* (ed. J.E. Smith and J.D. Lane), pp. 405–37. Elsevier, Amsterdam.

Wu, J.C. and Bunney, W.E. (1990). The biological basis of antidepressant response to sleep deprivation and relapse: review and hypothesis. *American Journal of Psychiatry*, **147**, 14–21.

Zisapel, N., Egozi, Y., and Laudon, M. (1982). Inhibition of dopamine release by melatonin: regional distribution in the rat brain. *Brain Research*, **246**, 161–3.

4

Disturbed hypothalamo–pituitary–adrenal axis regulation in depression: causes and consequences

I.N. FERRIER

INTRODUCTION

This chapter focuses on the hypothalamo–pituitary–adrenal (HPA) axis activity in depression. It is not intended as a comprehensive or exhaustive review of this topic but rather an overview of the pathophysiology of the HPA axis in depression and an attempt to describe the determinants and the consequences of this abnormality.

BACKGROUND

Overactivity of the HPA axis has been widely demonstrated in patients with primary affective disorder. Evidence for HPA activation in depression include reports of elevated plasma (Sachar et al. 1973), urinary (Carroll et al. 1976), and cerebrospinal fluid (Carroll et al. 1976) levels of corticosteroids and disruption of the 24 h cortisol secretory pattern (Sachar et al. 1973). In recent years, the most commonly employed test to examine HPA activity in depressed patients has been the dexamethasone suppression test (DST). Carroll and his co-workers (Carroll et al. 1981) reported that relative resistance to cortisol suppression by oral dexamethasone ('DST non-suppression') was found in a significant number of depressed patients. While the abnormalities in cortisol secretion and on the DST do not seem to be as specific to depression as was first thought (Arana et al. 1985; Braddock 1986) they are more common and more marked in depression than in other psychiatric disorders and within depressed patients DST non-suppression is more common in patients with melancholia and psychosis (Evans and Nemeroff 1987). Although cortisol is a stress-related hormone it appears that cortisol secretion in depressed patients is related to depression rather than anxiety and hypercortisolaemia is not a feature of anxiety states. Therefore it is suggested that elucidation of the mechanism of the hypercortisolaemia of depression may aid understanding of the determinants of depression—particularly as hypercortisolaemia could be a factor in the mechanism of the link between 'stress' and/or life events and depression.

MECHANISM OF HYPERCORTISOLAEMIA IN DEPRESSION

A schematic outline of the anatomy and physiology of the hypothalamo–pituitary–adrenal axis is shown in Fig. 4.1. There are many levels of homeostatic feedback in the axis and cortisol secretion is controlled through feedback at the pituitary and hypothalamic level. Glucocorticoid (GC) receptors are found throughout the brain: the highest density is found in hippocampus. Hippocampal GC receptors play a key role in the regulation of cortisol secretion particularly in the response to 'stress'.

In depression the secretion of cortisol is enhanced throughout the 24 h with an increased amplitude of secretory phases. Blunting of the normal diurnal rhythm of cortisol secretion is found in depression so that hyper-cortisolaemia is particularly evident in late evening and nocturnal samples. The changes seen in depression are largely state dependent although persistence of abnormality can be seen in some cases where it is a predictor of

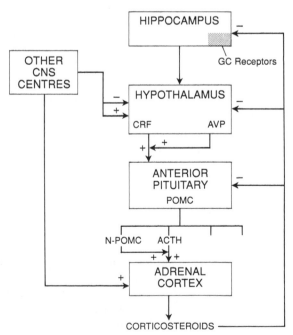

FIG. 4.1. Schematic representation of the limbic–hypothalamic–pituitary–adrenal axis. Abbreviations as in text: + = enhances, − = inhibits.

poor prognosis and increased frequency of relapse (Ferrier *et al.* 1991). The phenomena is more frequently seen in aged controls and elderly depressed patients, and in the latter group more time is needed for HPA axis normalization to occur (Greden *et al.* 1986; von Barbeleben and Holsboer 1991).

Adrenocorticotropic hormone (ACTH) levels are either normal or non-significantly elevated in major depression, although there have been conflicting reports. However, the cortisol response to ACTH appears to be enhanced in most studies of depressed patients whether the ACTH is derived exogenously or endogenously (for example following administration of corticotrophin releasing hormone (CRH)). The question of whether the threshold for the adrenal cortisol response to ACTH is altered in depression has not been resolved. These observations may link to the findings of Amsterdam *et al.* (1987) that adrenal gland volume (measured by CAT scans) is increased in depression. Since ACTH levels are normal in depression and in any event ACTH is involved in stimulating cortisol secretion rather than in inducing adrenal gland hypertrophy, the question remains as to the mechanism of enhanced adrenal gland size and sensitivity in depression. ACTH is derived from a larger parent molecule in the pituitary pro-opiomelanocortin (POMC) and the N-terminal end of POMC (N-POMC) enhances adrenal growth and induces hypertrophy. However our group in Newcastle (Charlton *et al.* 1988) was unable to demonstrate any dramatic change in N-POMC in depression and there is no evidence for failure of cosecretion of ACTH and N-POMC. These observations led us to speculate that the adrenal changes in depression were due to chronic relative overstimulation by ACTH and N-POMC in combination with enhanced activity in the neurogenic input to the adrenal cortex (Charlton and Ferrier 1989). This latter mechanism is a centrally driven process.

It is thought that ACTH levels in depression relate to a balance between the degree of negative feedback from cortisol and the degree of central drive from the hypothalamus mediated either through CRH or vasopression (AVP) or a combination of both. From a number of *in vitro* and *in vivo* studies, evidence has emerged that CRH and AVP have a synergistic effect on ACTH release from the pituitary (Gillies *et al.* 1984). There are numerous reports of blunted ACTH, but normal cortisol, responses to the administration of synthetic human CRH in depression. Initially it was thought that the blunted ACTH response to hCRH in depression was directly related to the hypercortisolaemic state since, in normal controls, circulating plasma cortisol determines the amount of ACTH released (Hermus *et al.* 1986) and a synthetic corticosteroid dexamethasone blunts the ACTH response to hCRH (von Bardleben *et al.* 1985). However it has been shown that a different response to dexamethasone suppression is found in depressed patients so that, in contrast to control subjects, there is substantial activation of the pituitary–adrenocortical system after adminis-tration of hCRH (Holsboer-Trachsler *et al.* 1991; von Bardleben and Holsboer 1991). These abnormalities are significantly positively related to both increasing age and increasing severity of depression. It has also been found that the size of the pituitary gland (measured by magnetic resonance imaging) is increased in depression and the change is more marked in elderly patients (Krishnan *et al.* 1991). These observations, taken together (and with others not detailed here) have led to the conclusions that (1) central drive from hypothamic CRH, vasopressin (and perhaps other secretagogues) is increased in depression (Charlton and Ferrier 1989; von Bardleben and Holsboer 1991), (2) that there are changes in the sensitivity of GC receptors in the brain to circulating glucocorticoids, and (3) both these processes are apparently accelerated by the ageing process.

Despite considerable progress in our understanding of the mechanism of hypercortisolaemia in depression the underlying cause(s) remain obscure. CRH secretion is increased by 'stress' but this is too simplistic a cause to account for the changes seen in depression. It is noteworthy that the CRF–ACTH–cortisol axis response to stress shows the phenomenon of 'priming' so that early or noxious stress for an animal is associated with an exaggerated response to a subsequent novel stressor (Thoman *et al.* 1968; Orr *et al.* 1990). Restrepo and Armario (1987) showed that chronically stressed animals had an exaggerated cortisol response to a novel stressor and that such animals had increased adrenal weight, an exaggerated ad-renal response to ACTH but normal basal levels of ACTH. These findings are analogous to the findings in depressed patients outlined above and may be linked to observations about the role and timing of stress in the aetiology of depression. The control of GC receptors in normal and depressed patients remains to be elucidated but again there is evidence from rat studies that early influences are important (Meaney *et al.* 1991). In rats, antidepres-sants increase GC receptor density and expression (Kitayama *et al.* 1988; Peiffer *et al.* 1991; Seckl *et al.*, 1992). The activation of GC receptors is required for the development of certain animal models of depression (Veldhuis *et al.* 1985) and blockade of GC receptors by the antagonist RU 38486 prevents their development (de Kloet *et al.* 1988).

The regulation of CRH secretion from the hypothalamus and the regula-tion of GC receptors in the brain are both related to the function of a variety of neurotransmitters. For example there is good evidence that serotonin (5-HT) plays a critical role both in the hypothalamus and in the limbic lobe (Spinedi and Gaillard 1991). However it is also clear that glucocorticoids have important effects on neurotransmitter function in-cluding inhibiting the crucial step between receptors and second messengers function (Lesch and Lerer 1991). The task for the next few years is to tease apart this 'chicken and egg' situation. To put the question in simple form, does abnormal 5-HT function in depression cause hypercortisolaemia or vice versa?

CONSEQUENCES OF HYPERCORTISOLAEMIA IN DEPRESSION

In the past few years increasing attention has been paid to the effects of hypercortisolaemia in depressed patients. However, it remains unclear if the hypercortisolaemic state of depressed patients has any specific consequences. It may be a marker for changes seen in some depressed patients—for example since raised steroids are more commonly found in the more severely depressed patients, hypercortisolaemia may be an 'innocent bystander' to some of the consequences of severe depression.

Depressed patients do not develop the stigmata of Cushing's syndrome (Gormley *et al.* 1985). This may be partly due to the fact that depressed patients do not exhibit raised levels of glucocorticoids for prolonged periods of time but this is not thought to be a sufficient reason. One possibility was that depressed patients did not have raised levels of free cortisol but our group have demonstrated that cortisol-binding globulin is unchanged in depression and that free cortisol is elevated substantially in many depressed patients (Leake *et al.* 1989). In studies of GC receptors in peripheral cells of depressed patients some groups found reductions in depressed, particularly hypercortisolaemic patients (Gormley *et al.* 1985; Whalley *et al.* 1986; Lowry *et al.* 1988) but others have not confirmed this (Schlecte and Sherman 1985; Rupprecht *et al.* 1991). Rupprecht *et al.* (1991) adduce evidence that in depressed patients there is partial steroid resistance (of uncertain cause) and this, for the moment, is where this approach must rest until newer investigative techniques are available.

There are a number of reports that hypercortisolaemia in depression is associated with poor outcome. In both younger and older depressed patients raised glucocorticoids have been associated with cognitive failure during an episode of depression (Rubinow *et al.* 1984; Siegel *et al.* 1989). Kellner *et al.* (1983) and Schlegel *et al.* (1989) showed that urinary and mean plasma cortisol respectively were positively correlated with elevated ventricular brain ratios (VBR) in younger populations of depressives. Results with the DST are more varied with Rao *et al.* (1989) finding a relationship between DST status and VBRs which was not confirmed by Targum *et al.* (1983) or Schlegel *et al.* (1989). We (Ferrier *et al.* 1991) recently demonstrated that DST nonsuppressor status of elderly depressed patients was associated with the development of cognitive decline and mild Parkinsonism over a 5-year follow-up period. These observations are reminiscent of Cushing's syndrome where there is evidence of depression, cognitive failure, and enlarged VBRs which are reversible following curative surgery.

The mechanism of enlarged VBRs and cognitive failure in depression is unknown and the site of any deleterious effect of glucocorticoids is similarly uncertain. Sapolsky (1987) put forward the hypothesis that elevated gluco-corticoids may induce selective hippocampal damage. Evidence from animal studies indicates that there are high densities of GC receptors in the hippo-campus (de Kloet and Reul 1987) and a similar distribution is found in human brain (Seckl et al. 1991). The ageing brain (including that of the human) undergoes a loss of hippocampal neurones (Miller et al. 1984) and, in the guinea pig, glucocorticoids in the upper physiological range increase the rate of senescent neurone loss in the hippocampus and increase the deleterious effects of other insults, for example hypoxia (McEwan et al. 1992). There is evidence of reduced hippocampal size in suicide victim's brains (Altshuler et al. 1990) and reduced temporal lobe volumes on MRI scans in resistant bipolar depressives (Altshuler et al. 1991). These reports require replication. The role of corticosteroids in normal ageing and psychiatric disease promises to be a major area of study during the next decade.

There is a considerable amount of research going on to ascertain whether the hypercortisolaemic state found in many depressives has any influence on the neurochemical, neuroendocrinological, or clinical changes seen in depression. Corticosteroids have effects on 5-HT, noradrenaline and doapmine function in rats (Nausieda et al. 1982; Dickinson et al. 1985) but whether the levels corticosteroids seen in depressed patients are related to changes in central monoamine function remains conjectural. Hypercortisolaemia in depression does not seem to influence the blunting of the TSH response to TRH but whether it affects the hormonal response to 5-HT agonists remains unclear. The lowering of plasma cortisol by metapyrone in patients with Cushing's syndrome is associated with an improvement in the severity of the associated depression (Jeffcoate et al. 1979; Kramlinger et al. 1985). A number of investigations are underway to elucidate whether metapyrone or the glucocorticoid receptor antagonist RU 38486 are of clinical benefit in depressed patients.

In summary, it remains a plausible hypothesis that corticosteroids may provide a neurobiological basis for the interaction between the genetic predisposition to depression and the effects of life events and stress (Bebbington and McGuffin 1989). The effects of age on the limbic–HPA axis and the response to stress are also of particular relevance to research in this area. The discovery of the hypercortisolaemia of depression led to the false hope that this was the biological marker of the condition. However, there is increasing evidence that continued research into the mechanism and consequence of hypercortisolaemia in depression may prove fruitful in terms of our understanding of the aetiology of depression and in the development of novel, perhaps, protective, therapies.

REFERENCES

Altshuler, L.L., Cassanova, M.F., Goldberg, T.E., and Kleinman, J.E. (1990). The hippocampus and parahippocampus in schizophrenia, suicide and control brains. *Archives of General Psychiatry*, **47**, 1029–34.

Altshuler, L.L., Conrad, A., Hauser, P., Ximing, L., Guze, B.H., Denkikoff, K. *et al.* (1991). Reduction of temporal lobe volume in bipolar disorder: a preliminary report of magnetic resonance imaging. *Archives of General Psychiatry*, **48**, 482–3.

Amsterdam, J.D., Marinelli, D.L., Arger, P., and Winokur, A. (1987). Assessment of adrenal gland volume by computed tomography in depressed patients and healthy volunteers: a pilot study. *Psychiatric Research*, **21**, 189 97.

Arana, G.W., Baldessarini, R.J., and Ornsteen, M. (1985). The dexamethasone suppression test for diagnosis and prognosis in psychiatry. Commentary and review. *Archives of General Psychiatry*, **42**, 1193–204.

Bebbington, P.E. and McGuffin, P. (1989). Interactive models of depression: the evidence. In *Depression, an integrative approach* (ed. K.R. Herbst and E.S. Paybel). Heinemann Medical, Oxford.

Braddock, L. (1986). The dexamethasone suppression test: fact or artefact. *British Journal of Psychiatry*, **148**, 363–74.

Carroll, B.J., Curtis, G.C., Davis, B.M., Mendels, J., and Sugerman, A. (1976). Urinary-free cortisol excretion in depression. *Psychological Medicine*, **6**, 43–50.

Carroll, B.J., Feinberg, M., Greden, J., Tarika, J., Albala, A.A., Haskett, R.F. *et al.* (1981). A specific laboratory test for the diagnosis of melancholia. *Archives of General Psychiatry*, **38**, 15–22.

Charlton, B.G. and Ferrier, I.N. (1989). Hypothalamo–pituitary–adrenal axis abnormalities in depression: a review and a model. *Psychological Medicine*, **19**, 331–6.

Charlton, G.B., Ferrier, I.N., Leake, A., Edwardson, J.A., Eccleston, D., Crowcombe, K. *et al.* (1988). A multiple time point study of N-terminal pro-opiomelanocortin in depression using a two-site recognition immunoradiometric assay. *Clinical Endocrinology*, **28**, 165–72.

de Kloet, E.H. and Reul, J.M.H.M. (1987). Feedback action and tonic influences of corticosteroids on brain function: a concept arising from heterogeneity of brain receptor systems. *Psychoneuroendocrinology*, **12**, 83–105.

de Kloet, E.R., de Kock, S., and Schild, V. (1988). Anteglucocorticoid RU38486 attenuates retention of a behaviour and disinhibits the hypothalamic–pituitary and adrenal axis at different brain sites. *Neuroendocrinology*, **47**, 109–15.

Dickinson, S.L., Kennett, G.A., and Curzon, G. (1985). Reduced 5-hydroxy-tryptamine-dependent behaviour in rats following chronic corticosterone treatment. *Brain Research*, **345**, 18–19.

Evans, D.L. and Nemeroff, C.B. (1987). The clinical use of the dexamethasone suppression test in DSM—III affective disorders: correlation with the severe depressive subtypes of melancholia and psychosis. *Journal of Psychiatric Research*, **21**, 185–94.

Ferrier, I.N., Lister, E.S., Riordan, P.M., Scott, J.L., Lett, D.J., Leake, A. *et al.* (1991). A follow-up study of elderly depressives and Alzheimer-type dementia—relationship with DST status. *International Journal of Geriatric Psychiatry*, **6**, 279–86.

Gillies, G., Puri, A., Hodgkinson, S., and Lowry, P.J. (1984). Involvement of rat corticotrophin-releasing factor-41-related peptide and vasopressin in adrenocorticotrophin-releasing activity from superfused rat hypothalami *in vitro*. *Journal of Endocrinology*, **103**, 25–9.

Gormley, G.J., Lowry, M.T., Reder, A.T., Hoselhorn, V.D., Antel, J.P., and Metzer, H.Y. (1985). Glucocorticoid receptors in depression. Relationship to the dexamethasone suppression test. *American Journal of Psychiatry*, **142**, 1278–84.

Greden, J., Flegel, P., Haskett, R., Dilsaver, S., Caroll, B., and Grunhaus, L. (1986). Age effects in serial hypothalamic–pituitary–adrenal monitoring. *Psychoneuroendocrinology*, **11**, 195–203.

Hermus, A.R., Pieters, G.F., Posman, G.J., Hofman, J., Smals, A.G., Benrad, T.J. *et al.* (1986). Escape from dexamethasone-induced ACTH and cortisol suppression by corticotrophin-releasing hormone: modulatory effect of basal dexamethasone levels. *Clinical Endocrinology*, **26**, 67–74.

Holsboer-Trachsler, E., Stohler, R., and Hatzinger, M. (1991). Repeated administration of the combined dexamethasone-human corticotrophin releasing hormone stimulation test during treatment of depression. *Psychiatry Research*, **38**, 163–71.

Jeffcoate, W.J., Silverstone, J.T., Edwards, C.W.R., and Besser, G.M. (1979). Psychiatric manifestations of Cushing's syndrome: response to lowering of plasma cortisol. *Quarterly Journal of Medicine*, **48**, 465–72.

Kellner, C.H., Rubinow, D.R., Gold, P.W., and Post, R.M. (1983). Relationship of cortisol hypersecretion to brain CT scan alterations in depressed patients. *Psychiatric Research*, **8**, 191–7.

Kitayami, I., Cintra, A.M., Fuxe, K., Agnati, L.F., Oegren, S.O., Härfstrand, A. *et al.* (1988). Effects of chronic imipramine treatment on glucocorticoid receptor immunoreactivity in various regions of the rat brain. *Journal of Neural Transmission*, **73**, 191–203.

Kramlinger, K.G., Peterson, G.C., and Watson, P.K. (1985). Metyrapone for depression and delirium secondary to Cushing's syndrome. *Psychosomatics*, **26**, 67–71.

Krishnan, K.R.R., Doraiswamy, P.M., Lurie, S.N., Figiel, G.S., Husein, M.M., Boyko, O.B. *et al.* (1991). Pituitary size in depression. *Journal of Clinical Endocrinology and Metabolism*, **72**, 256–9.

Leake, A., Griffiths, H.W., Pascual, J.A., and Ferrier, I.N. (1989). Corticosteroid-binding globulin in depression. *Clinical Endocrinology*, **30**, 39–45.

Lesch, K.P. and Lerer, B. (1991). The 5-HT receptor-G-protein-effector system complex in depression. I. Effect of glucocorticoids. *Journal of Neural Transmission*, **84**, 3–18.

Lowry, M.T., Reder, A.T., Gormley, G.T., and Meltzer, H.Y. (1988). Comparison of *in vivo* and *in vitro* glucocorticoid sensitivity in depression: relationship to the dexamethasone suppression test. *Biological Psychiatry*, **24**, 619–30.

McEwan, B.S., Gould, E.A., and Sakai, R.R. (1992). The vulnerability of the hippocampus to protective and destructive effects of glucocorticoids in relation to stress. *British Journal of Psychiatry*, **160** (Suppl. 15), 18–24.

Meaney, M.J., Mitchell, J.B., Aitken, D.H., Bhatnagor, S., Bodnoff, S.R., Iny, L.J. *et al.* (1991). The effects of neonatal handling on the development of the adrenocortical response to stress. Implications for neuropathology and cognitive deficits in later life. *Psychoneuroendocrinology*, **16**, 85–103.

Miller, A.K., Alston, R.L., Mountjoy, C.Q., and Corsellis, J.A. (1984). Automated differential cell counting on a sector of the normal human hippocampus: the influence of age. *Neuropathology and Applied Neurobiology*, **10**, 123–41.

Nausieda, P.A., Carvey, P.M., and Weiner, W.J. (1982). Modification of central serotonergic and dopaminergic behaviours in the course of chronic corticosteroid administration. *European Journal of Pharmacology*, **78**, 335–43.

Orr, T.E., Meyerhoff, J.L., Mougey, E.H., and Bunnell, B.N. (1990). Hyperresponsiveness of the rat neuroendocrine system due to repeated exposure to stress. *Psychoneuroendocrinology*, **15**, 317–28.

Peiffer, A., Veilleux, S., and Barden, N. (1991). Antidepressant and other centrally acting drugs regulate glucocorticoid receptor messenger RNA levels in rat brain. *Psychoneuroendocrinology*, **16**, 505–15.

Rao, P.V., Krishnan, K.R.R., Goli, V., Saunders, W.B., Ellinwood, E.H., Blazer, D.G. *et al.* (1989). Neuroanatomical changes and hypothalamo–pituitary–adrenal axis abnormalities. *Biological Psychiatry*, **26**, 729–32.

Restrepo, C. and Armario, A. (1987). Chronic stress alters pituitary–adrenal function in prepubertal rats. *Psychoneuroendocrinology*, **12**, 393–8.

Rubinow, H.R., Post, R.M., Savard, R., and Gold, P.W. (1984). Cortisol hypersecretion and cognitive impairment in depression. *Archives of General Psychiatry*, **41**, 279–83.

Rupprecht, R., Kornhuber, J., Norbert, W., Lugauer, J., Göbel, C., Haack, D. *et al.* (1991). Disturbed glucocorticoid receptor autoregulation and corticotrophin response to dexamethasone in depressives pretreated with metyrapone. *Biological Psychiatry*, **29**, 1099–109.

Sachar, E.J., Hellman, L., Roffuarg, H.P., Halpern, F.S., Fukushima, D.K., and Gallaher, R.F. (1973). Disrupted 24-hour patterns of cortisol secretion in psychiatric depression. *Archives of General Psychiatry*, **28**, 19–21.

Sapolsky, R.M. (1987). Glucocorticoids and hippocampal damage. *Trends Neurosci.* **10**, 346–9.

Schlechte, J.A. and Sherman, B. (1985). Lymphocyte glucocorticoid receptor binding in depressed patients with hypercortisolemia. *Psychoneuroendocrinology*, **10**, 469–74.

Schlegel, S., Von Bardeleben, U., Wiedemann, K., Frommberger, U., and Holsboer, F. (1989). Computerized brain tomography measures compared with spontaneous and suppressed plasma cortisol levels in major depression. *Psychoneuroendocrinology*, **14**, 209–16.

Seckl, J.R. and Fink, G. (1992). Antidepressants increase glucocorticoid and mineralocorticoid receptor mRNA expression in rat hippocampus *in vivo*. *Neuroendocrinology*, **55**, 621–6.

Seckl, J.R., Dickson, K.L., Yates, C., and Fink, G. (1991). Distribution of gluco-corticoid and mineralocorticoid receptor messenger RNA expression in human postmortem hippocampus. *Brain Research*, **561**, 332–7.

Siegel, B., Gurevich, D., and Oxenbrug, G.F. (1989). Cognitive impairment and cortisol resistance to dexamethasone suppression in elderly depression. *Biological Psychiatry*, **25**, 229–34.

Spinedi, E. and Gaillard, R.C. (1991). Stimulation of the hypothalamo–pituitary–adrenocortical axis by the central serotonergic pathway: involvement of endogenous corticotropin-releasing hormone but not vasopressin. *Journal of Endocrinological Investigation*, **14**, 551–7.

Targum, S.D., Rosen, L.N., Delisi, L.E., Wernberger, D.R., and Citrin, C.M. (1983). Cerebral ventricular size in major depressive disorder: association with delusional symptoms. *Biol. Psychiat.*, **18**, 329–36.

Thoman, E.B., Levine, S., and Arnold, W.J. (1968). Effects of maternal deprivation and incubator rearing on adrenocortical activity in the adult rat. *Developmental Psychobiology*, **1**, 21–3.

Veldhuis, H.D., De Korte, C.C.M.M., and De Kloet, E.R. (1985). Glucocorticoids facilitate the retention of acquired immobility during forced swimming. *European Journal of Pharmacology*, **115**, 211–17.

von Bardeleben, U. and Holsboer, F. (1991). Effect of age on the cortisol response to human corticotrophin-releasing hormone in depressed patients pretreated with dexamethasone. *Biological Psychiatry*, **29**, 1042–50.

von Bardeleben U., Holsboer, F., Stalla, G.K., and Müller, O.A. (1985). Combined administration of human corticotrophin-releasing factor and lysine vasopressin induces cortisol escape from dexamethasone suppression in healthy subjects. *Life Sciences*, **37**, 1613–18.

Whalley, L.J., Borthwick, N., Copolov, D., Dick, H., Christie, J.E., and Fink, G. (1986). Glucocorticoid receptors and depression. *British Medical Journal*, **292**, 859–61.

5

Hormones, genes, and the triggering of bipolar illness in the puerperium

S.A. CHECKLEY, A. WIECK, J.A. BEARN, M.N. MARKS, I.C. CAMPBELL, and R. KUMAR

Childbirth is a powerful trigger for the relapse of bipolar affective disorder and so provides a convenient experimental paradigm for the investigation of mechanisms responsible for the triggering of affective illness. 50 per cent of all patients with bipolar affective disorders may relapse in the 6 weeks following childbirth (Dean *et al*. 1989). There is no increased incidence for the first 4 days postpartum but there is subsequently a dramatic increase in the rate of onset of psychosis over the next 10 days (Brockington *et al*. 1982). This time course is quite different to that which has been reported for the onset of nonpsychotic depressive disorder which has a much wider spread of time of postnatal onset with a 'peak' at 6–12 weeks after delivery (O'Hara and Zekoski 1988). This precise relationship between the time of parturition and the time of the relapse of bipolar (as compared to unipolar) illness is an important clue to the presence of a biological triggering process.

A second clue arises from the magnitude of the effect of childbirth on relapse. Life events result in a 2–7fold increase in the rate of relapse of unipolar depression (Paykell 1978) and the increase in rate of unipolar depression following childbirth is of a similar order. However, childbirth is followed by a thirtyfold increase in the rate of bipolar relapse (Kendell *et al*. 1987) and in patients with a previous puerperal illness, the risk is very much greater (Dean *et al*. 1989).

The relapse of a small proportion of puerperal illness with the onset of the first menstrual period (Brockington *et al*. 1989) suggests a biological mechanism and possibly one which involves sex steroids.

ENDOCRINE CHANGES IN THE POSTNATAL PERIOD

The endocrine system is the most likely biological link between the brain and the biological mechanisms concerned with childbirth. It is therefore appropriate to consider the possible involvement of hormones secreted by the mother, by the placenta and by the placenta and fetus acting together as the 'fetoplacental unit'.

Maternal hormones

Most of the changes in maternal hormones in the puerperium are a function of childbirth itself. Oxytocin is secreted to augment contractility and stress sensitive hormones such as beta endorphin are secreted in response to the trauma of labour (Wieck 1990 for review). However, these changes are unlikely to be the trigger for psychosis because such changes do not accompany caesarian section which is followed by puerperal psychosis just as frequently as in childbirth by vaginal delivery (Kendell *et al.* 1987). Similarly the endocrinology of lactation is unlikely to be the trigger for puerperal psychosis since in the first 4 days postpartum lactation develops spontaneously in all mothers regardless of whether or not they subsequently suppress lactation: psychoses develop from the fifth day postpartum before the suppression of lactation in some women. Furthermore neither oxytocin nor prolactin freely cross the blood brain barrier.

Placental hormones

The placenta synthesizes a wide range of peptide growth factors including insulin—like growth factor -I (IGF-I), epidermal growth factor (EGF), and transforming growth factor alpha (TGF»ga) (Bonney and Franks 1990). The placenta also synthesizes the glycoprotein hormones, placental lactogen, and human chorionic gonadotrophin. However none of these hormones is likely to trigger bipolar relapse since none cross the blood brain barrier to a significant extent.

Fetoplacental hormones

The fetus and placenta act as a functional unit which by the end of pregnancy synthesizes virtually all of the maternal sex steroids. Dihydroepiandosterone (DHEAS) is synthesized in the fetal zone of the adrenal gland of the fetus and is subsequently metabolized to oestrone and oestradiol within the placenta. Similarly 16 alpha hydroxy dihydroepiandosterone (16 OH DHEAS) is synthesized in the fetal liver and is subsequently metabolized to oestriol in the placenta. The synthesis of progesterone is simpler and by the end of pregnancy mostly occurs within the placenta. This means that when the fetus and placenta are delivered virtually all synthesis of sex steroids within the mother ceases. Plasma concentrations of oestrogen and progesterone which have risen by two orders of magnitude during pregnancy, fall by that amount over 48 h (Cinque *et al.* 1985; Cowley and Mason 1990).

These changes will be detected by oestrogen and progesterone receptors within maternal limbic forebrain regions such as the hippocampus, amygdala, and hypothalamus (Greenstein 1986). Steroid receptor complexes at these sites bind to DNA and modify gene expression (Beato 1989) and the resulting change in the synthesis of peptides occurs with a time scale of days or weeks. Thus, the rapid withdrawal of sex steroids from maternal brain will result in changes in the rates of synthesis of a number of proteins throughout the limbic forebrain with a time course which matches the onset of bipolar relapses. The next section will look at a neuroendocrine model of central oestrogen receptor-mediated function in the human post-natal period, and the final section will look at the possible changes in central dopamine function which may result.

THE OESTROGEN-SENSITIVE NEUROPHYSIN (ESN) TEST

Oestrogen-sensitive neurophysin (ESN) is the carrier protein for the posterior pituitary hormone oxytocin. The precursor preprooxytocin (Robinson 1986) is synthesized in the anterior hypothalamus in cells which contain oestrogen receptors (Sar and Stumpf 1980). In the presence of oestrogen, preprooxytocin is synthesized in the anterior hypothalamus, transported to the posterior pituitary, cleaved, and then secreted as oxytocin and ESN. Oxtyocin is rapidly metabolized but ESN can conveniently be measured in plasma. Following the administration of 20–200 μg ethinyl oestradiol, plasma concentrations of ESN are raised in healthy female volunteers in a dose-dependent manner. The time course of this effect is notable. Whereas most neuroendocrine responses take place within minutes, the time course of the ESN response to ethinyl oestradiol is measured in days. This time course is consistent with a genomic action of ethinyl oestradiol. We have studied the effect of the postnatal period on the ESN response to ethinyl oestradiol. The response is reduced throughout the first 4 weeks post-partum. In control experiments, we have reported normal ESN responses to ethinyl oestradiol in patients with anorexia nervosa who were matched with the postpartum women for low plasma oestrogen concentrations (Bearn *et al.* 1990) and in menopausal women (unpublished data) with similarly low plasma oestrogen concentrations. This suggests that the reduced ESN response in the postpartum women is not simply a function of the ambient oestradiol concentrations. More likely, however, it reflects a dynamic effect of oestrogen withdrawal on oestrogen receptors. Thus in mice, reduced numbers of oestrogen receptors have been found in the anterior hypothalamus in the postnatal period (Koch and Ehret 1989): such a reduction in oestrogen receptors would explain our finding of reduced ESN responses in the postnatal period.

To date we have found a reduction in a neuroendocrine measure of the responsiveness of central oestrogen receptors in the postnatal period. A number of experimental questions can now be asked to test the relationship between oestrogen receptor function and the triggering of bipolar affective disorder.

1. Does the size of the ESN response to ethinyl oestradiol predict relapse of bipolar illness?
2. Does the size or rate of the postpartum fall in plasma oestradiol predict relapse?
3. Can treatment with oestradiol immediately after childbirth prevent the subsequent relapse of affective illness?

CENTRAL DOPAMINE RECEPTOR FUNCTION IN THE POSTPARTUM PERIOD

The evidence which has been reviewed thus far suggests that the withdrawal of oestrogen (or possibly progesterone) can trigger bipolar relapse presumably by genomic action. It seems unlikely that the steroids exert this effect by direct action at any manic depressive gene or genes. If this were so, then one might predict a relapse after every childbirth if not after every menstrual cycle. An indirect link between sex steroids and bipolar affective illness would more closely fit the clinical facts. The following data suggests that dopamine might provide the link.

1. The psychotic disorders which are common in the postnatal period are ones in which dopamine is apparently involved. Thus mania and schizomania can be treated by dopamine receptor antagonists and mimicked by amphetamine.

2. Bipolar depression may be accompanied by neuroendocrine evidence of enhanced dopamine-mediated responses (Gold *et al*. 1974). Bipolar disorder is prevented by lithium, the actions of which include the stabilization of dopamine receptors (Pittman *et al*. 1984). Thus a reasonable case can be deduced from the literature to implicate enhanced dopamine-mediated function in the triggering of bipolar affective disorder in the postpartum period.

3. The animal literature provides complementary evidence in so far as the rapid withdrawal of high doses of oestrogens can result in enhanced dopamine-mediated behavioural responses (Gordon and Perry 1983).

4. Finally there is anecdotal clinical evidence that tardive dyskinesia can be exacerbated following childbirth (Vinogradov and Csernansky 1990), which would suggest enhanced dopaminergic neurotransmission in the postnatal period.

For these reasons we and others (Cookson 1985) have argued that sex-steroid withdrawal triggers the onset of bipolar affective disorder in the puerperium by sensitizing central dopamine receptors. We have tested this hypothesis in the following way.

GROWTH HORMONE RESPONSES TO APOMORPHINE IN THE POSTNATAL PERIOD IN HEALTHY CONTROL AND HIGH-RISK PATIENTS

The growth hormone (GH) response to apomorphine depends on the stimulation of central dopamine D_2 receptors (Lal *et al.* 1990). If puerperal psychosis is triggered by the sensitization of central dopamine receptors, then this sensitization should result in enhancement of the GH response to apomorphine, particularly in those patients who are about to develop bipolar relapses.

To test this we have measured GH responses to apomorphine on the fourth day postpartum in 15 healthy controls and 15 high-risk patients who had suffered from previous bipolar affective illnesses. None of the patients had general anaesthetics at childbirth and none had received psychotropic medication throughout pregnancy.

The high-risk patients had larger GH responses than did the controls, and within the high-risk patients those who subsequently relapsed had larger GH responses than the high-risk patients who remained well (Wieck *et al.*, 1991).

The high-risk patients who did not relapse had very similar responses to the healthy controls and also to seven puerperal women with a past history of unipolar depression. These data support the hypothesis that sensitization of dopamine receptors is involved in the triggering of psychosis.

The same hypotheses also predicts lower resting plasma prolactin concentrations in the high-risk relapses as compared to the controls and in the high-risk patients who subsequently do not relapse.

Thus, we have evidence for the sensitization of central dopamine receptors in neuroendocrine systems immediately prior to the relapse of manic and schizomanic illnesses in the postnatal period. It remains to be established whether or not enhanced dopamine-mediated behaviour can be detected in other aspects of brain function prior to the development of puerperal relapse. It also remains to be seen whether or not neuroleptic drugs and/or lithium treatment might block the sensitization of central dopamine receptors in the postnatal period and whether or not these same treatments might also prevent the development of puerperal psychosis in high-risk individuals.

OVERALL CONCLUSIONS

Our clinical studies support the hypothesis that sex-steroid withdrawal triggers bipolar affective disorder in the postnatal period by the sensitization of central dopamine receptors. A number of ongoing studies to further test this hypothesis have been described. Some broader issues are now briefly discussed.

Schizophrenia

No comparable data has been presented on the relapse of schizophrenia in the postpartum period. It would be interesting to test the same hypothesis although a different method would be needed as many patients would require neuroleptic treatment throughout pregnancy and this would preclude measurement of the GH response to apomorphine. One could, however, measure the effects of neuroleptics on prolactin and on tremor and predict that at constant dose, these drug effects would be *reduced* prior to relapse as an effect of the sensitization of central dopamine receptors.

Unipolar depression

There are no data to suggest that the sensitization of dopamine receptors is related to the onset of unipolar depression in the postnatal period. We have found normal GH responses to apomorphine in these patients and as we have already discussed the onset to unipolar depressions is not as closely linked to the time of childbirth as are the psychoses. Indeed, for the following reasons psychosocial factors can be implicated.

Psychosocial factors

In a study of 45 controls, 26 bipolar, and 17 unipolar pregnant women, we (Marks *et al.* 1992) have found that unipolar depression following depression could be predicted by a combination of a past history of unipolar or bipolar depression and a combination of life events, neuroticism, and social difficulties (Marks *et al.* 1992). None of these social factors predicted psychosis in bipolar patients, although they did predict nonpsychotic relapse in patients. The only psychosocial predictor of psychotic (manic, schizomanic, or delusional depressive) relapse in the bipolar patients was a low expressed emotion in the husband's behaviour to the wife's. Psychotic relapse was also predicted by a short interval from the last episode and, as there was an intercorrelation between this measure and low expressed emotion in the husband, it is possible that the husband's behaviour is a reaction to the wife's vulnerability to illness rather than being the cause of it.

Sleep disorder and the onset of bipolar affective disorder

Sleep disorder is a common precursor of hypomania and a common symptom of the minor mood disorders ('the fourth day blues') which are seen in half of all puerperal women. Amphetamine disturbs sleep and it is possible that dopamine could be involved in the sleep disorder prior to the onset of puerperal psychosis: this too might predict bipolar relapse.

Bipolar patients with and without puerperal relapses

Although 50 per cent of patients with a previous puerperal illness may develop a psychotic relapse following each subsequent pregnancy, it seems most improbable that half of all female bipolar patients will have relapsed following each pregnancy. A direct test of this possibility is needed but if confirmed, would point to a biological subdivision of the bipolar affective disorder category. Furthermore, since the mode of action of steroid hormones is likely to be genomic, there may well be genetic and molecular genetic differences between bipolar patients with and without puerperal relapse. Individual patients with puerperal relapses of bipolar illness report premenstrual depression which may also help to distinguish the two groups of bipolar patients.

The triggering of bipolar affective illness outside the puerperium

If bipolar relapses in the puerperium involve the sensitization of dopamine receptors then dopamine might be involved in the triggering of bipolar relapses at other times in the life cycle. Indeed, lithium stabilizes dopamine receptors following changes in dopamine availability (Pittman *et al.* 1984) and may prevent bipolar relapses in this way. In healthy volunteers, lithium may not reduce the GH response to apomorphine (Lal *et al.* 1978) but in situations where the GH response to apomorphine changes such as across the menstrual cycle and in the puerperium, then lithium might stabilize this dopamine receptor-mediated neuroendocrine response.

Prevention of bipolar relapses in the puerperal

The hypothesis that oestrogen withdrawal triggers bipolar relapse by sensitization of dopamine receptors predicts that each of three preventative treatments should be effective. The first is oestrogen administration. If a dose of oestrogen can be given which slows the rate of oestrogen withdrawal without causing untoward effects, for example on the coagulation system, then such a treatment should prevent relapse. Secondly, if lithium treatment prevents change in the GH response to apomorphine in the puerperium then lithium should prevent bipolar relapse in the puerperium. Thirdly, neuroleptic treatment initiated within 4 days of childbirth should in adequate doses prevent relapse. The long term aims of this research is to prevent relapse and these three clinical strategies are important both from a theoretical point of view and hopefully in the prevention of puerperal psychosis.

REFERENCES

Bearn, J.A., Fairhall, K.M., Robinson, I.C.A.F., Lightman, S.L., and Checkley, S.A. (1990). Changes in a proposed neuroendocrine marker of central oestrogen receptor function in postpartum women. *Psychological Medicine*, **20**, 779–84.
Beato, M. (1989). Gene regulation by steroid hormones. *Cell*, **56**, 335–44.
Brockington, I.F., Winokur, G., and Dean, C. (1982). Puerperal Psychosis. In *Motherhood and mental illness* (eds. I.F. Brockington and R. Kumar), pp. 37–69. Academic Press, London.
Brockington, I.F., Kelly, A., Hall, P., and Deakin, W. (1988). Premenstrual relapse of puerperal psychosis. *Journal of Affective Disorders*, **14**, 287–92.
Checkley, S.A. (1991). Neuroendocrine mechanisms and the precipitation of depression by life events. *British Journal of Psychiatry*, **16**, (Suppl. 15), 7–17.
Cinque, B., Montesanti, M.I., Parlati, E., Cucisano, A., Montemurro, A., Maniccia, E. *et al.* (1985). Corpus luteum function during the early puerperium. *Journal of Endocrinological Investigations*, **8**, 1–6.
Conley, A.J. and Mason, J.I. (1990). Placental steroid hormones. *Baillières Clin. Endocrinology and Metabolism*, **4**, 249–72.
Costain, D.W., Gelder, M.G., Cowen, P.J., and Grahame-Smith, D.G. (1982). Electroconvulsive therapy and the brain: evidence for increased dopamine-mediated responses. *Lancet*, **ii**, 400–4.
Dean, C., Williams, R.J., and Brockington, I.F. (1989). Is puerperal psychosis the same as bipolar manic depressive disorder? A family study. *Psychological Medicine*, **19**, 637–47.
Gold, P.M., Goodwin, F.K., and Wher, J.A. (1976). Growth hormone and prolactin responses to L-DDPA in affective illness. *Lancet*, **ii**, 1308–9.
Gordon, J.H. and Perry, K.D. (1983). Pre- and postsynaptic neurochemical alterations following oestrogen-induced striatal dopamine hypo- and hypersensitivity. *Brain Research Bulletin*, **10**, 425–8.
Greenstien, B. (1986). Steroid hormone receptors in the brain. In *Neuroendocrinology* (ed. S.L. Lightman and D.J. Everitt), pp. 32–48. Blackwell Scientific, Oxford.
Kendell, R.E., Chalmers, J.C., and Platz, C. (1987). Epidemiology of puerperal psychosis. *British Journal of Psychiatry*, **150**, 662–73.
Koch, M. and Ehret, G. (1989). Immunocytochemical localisation and quantitation

of oestrogen-binding cells in the male and female (virgins, pregnant lactating) mouse brain. *Brain Research*, **489**, 101–12.

Lal, S., Nair, V.P.V., and Guyda, H. (1978). Effect of lithium on hypothalamic-pituitary-dopaminergic function. *Acta Psychiatrica Scandinavica*, **57**, 91–6.

Lal, S., Nair, V., Thavundayil, N.P., Tawar, V., Quirion, R., and Guyda, H. (1991). Stereospecificity of dopamine receptor mediating the growth hormone response to apomorphine in man. *Journal of Neural Transmission*, **85**, 157–67.

Marks, M.N., Wieck, A., Checkley, S.A., and Kumar, R. (1992). Contribution of psychological and social factors to psychotic and non-psychotic relapse after childbirth in women with previous histories of affective disorder. *Journal of Affective Disorders*, (In press.)

Paykell, F.S. (1978). Contribution of life events to the causation of psychiatric illness. *Psychological Medicine*, **8**, 245–53.

Pittman, K.J., Jakubovic, A., and Fibiger, H.C. (1984). The effects of chronic lithium on behavioural and biochemical indices of dopamine receptor supersensitivity in the rat. *Psychopharmacology*, **82**, 371–7.

Robinson, I.C.A.F. (1986). The magnocellular and parvocellular OT and AVP systems. In *Neuroendocrinology* (ed. J.L. Lightman and B.J. Everitt), pp. 154–76. Blackwell Scientific, Oxford.

Sar, M. and Stumpf, W.E. (1980). Simultaneous localisation of 3-H oestradiol and neurophysin on arginine vasopressin in hypothalamic neurons demonstrated by a combined technique of dry mount autoradiography and immunohistochemistry. *Neuroscience Letters*, **17**, 179–84.

Vinogradov, S. and Cseruansky, J.C. (1990). Postpartum psychosis with abnormal movements: dopamine supersensitivity unmasked by withdrawal of endogenous oestrogens. *Journal of Clinical Psychiatry*, **51**, 365–6.

Watson, J.P., Elliott, S.A., Rugg, A.J., and Brough, D.I. (1984). Psychiatric disorder in pregnancy and the first postnatal year. *British Journal of Psychiatry*, **144**, 453–62.

Wieck, A. (1989). Endocrine aspects of postnatal mental disorders. In *Psychological aspects of obstetrics and gynaecology*, Vol. 3/4, pp. 857–77. Baillière Tindall, London.

Wieck, A., Kumar, R., Hirst, A.D., Marks, M.N., Campbell, I.C., and Checkley, S.A. (1991). Increased sensitivity of dopamine receptors and recurrence of affective psychosis after childbirth. *British Medical Journal*, **303**, 613–16.

6

5-HT$_{1A}$ receptors and antidepressant drug action

S.E. GARTSIDE and P.J. COWEN

INTRODUCTION

Experimental studies suggest that many clinically useful antidepressant drug treatments have profound effects on brain serotonin (5-hydroxy-tryptamine; 5-HT) function. Particular attention has been focused on the changes that antidepressant treatments can produce in the sensitivity of brain 5-HT receptors since the time-course of these adaptive changes often parallels the onset of clinical activity of drug therapies.

The recent description of multiple 5-HT receptor subtypes has given a new impetus to investigations of the effect of antidepressant treatments on brain 5-HT receptor populations. It is now clear that 5-HT pathways interact with a variety of 5-HT receptor subtypes which differ in their anatomical localization, functional correlates, and adaptive response to antidepressant treatment. The 5-HT$_{1A}$ receptor is of particular interest in relation to antidepressant drug treatment because of the role of this receptor subtype, both as an autoreceptor controlling the activity of 5-HT neurones, and as a postsynaptic receptor mediating the effects of released 5-HT. In addition, recent clinical evidence suggests that selective 5-HT$_{1A}$ receptor ligands have anxiolytic and antidepressant properties (Knapp 1985; Jenkins *et al*. 1990).

This chapter reviews the effect of various antidepressant treatments on models of 5-HT$_{1A}$ autoreceptor and postsynaptic 5-HT$_{1A}$ receptor function in animals and humans. The effects that such changes in 5-HT$_{1A}$ receptor sensitivity may have on overall postsynaptic 5-HT$_{1A}$ receptor-mediated neurotransmission are discussed.

While many of these investigations, particularly those involving human subjects, are at an early stage, the present findings suggest that several effective antidepressant drug treatments increase overall neurotransmission through postsynaptic 5-HT$_{1A}$ receptor synapses, but that the way in which this is achieved varies according to the drug being studied.

5-HT$_{1A}$ RECEPTORS

Localization and biochemical properties

The 5-HT$_{1A}$ receptor is a member of the 5-HT$_1$ receptor family which have in common a very high affinity for 5-HT, and a G protein-linked second messenger system. 5-HT$_{1A}$ receptors can be selectively labelled by [^{3}H]-8-hydroxy-2-(di-*n*-propylamino) tetralin (8-OH-DPAT), a 5-HT$_{1A}$ receptor agonist. Autoradiographic studies in rat brain using [^{3}H]-8-OH-DPAT reveal a high density of 5-HT$_{1A}$ binding sites in both the midbrain raphe nuclei, and in certain forebrain regions including hippocampus, hypothalamus, and entorhinal and frontal cortex (Marcinkiewicz *et al.* 1984; Pazos and Palacios 1985; Weissmann-Nanopoulos *et al.* 1985). In the raphe nuclei 5-HT$_{1A}$ binding sites are located on the cell bodies and/or dendrites of 5-HT neurones, whilst in forebrain regions, 5-HT$_{1A}$ binding sites are found postsynaptically to 5-HT neurones (Hall *et al.* 1985; Vergé *et al.* 1985) (Fig. 6.1). A similar distribution of 5-HT$_{1A}$ binding sites has now been reported in other mammalian species including man (Hoyer *et al.* 1986; Pazos *et al.* 1987).

The 5-HT$_{1A}$ receptor is negatively coupled to adenylate cyclase via a G protein. Thus, in hippocampal membranes for example, 5-HT is reported to inhibit forskolin-stimulated adenylate cyclase activity through activation of 5-HT$_{1A}$ receptors (De Vivo and Maayani 1986; Bockaert *et al.* 1987). 5-HT$_{1A}$ receptors in both hippocampus and dorsal raphe nucleus have been shown to be linked via a G protein to potassium channels, such that, independently of effects on adenylate cyclase, 5-HT$_{1A}$ receptors mediate membrane hyperpolarization via the opening of potassium channels (Andrade and Nicoll 1987; Innis *et al.* 1988).

5-HT$_{1A}$ receptors located on 5-HT nerve cell bodies and/or dendrites in the raphe nuclei, function as autoreceptors. Thus, stimulation of 5-HT$_{1A}$ receptors in the raphe nuclei inhibits the firing of 5-HT neurones (Hjorth

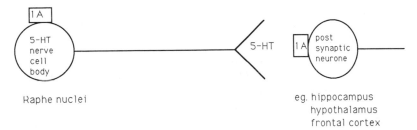

FIG. 6.1. Anatomical location of 5-HT$_{1A}$ receptors in mammalian brain. 5-HT$_{1A}$ receptors are found on 5-HT cell bodies where they function as inhibitory autoreceptors and at sites postsynaptic to 5-HT neurones in certain brain regions.

et al. 1982; Sprouse and Aghajanian 1986), and causes a decrease in the release of 5-HT in terminal fields (Sharp *et al.* 1989*a,b*).

Functional correlates of 5-HT$_{1A}$ receptor activation

Stimulation of 5-HT$_{1A}$ receptors in the rat produces a number of functional effects, including alterations in motor activity (Hjorth *et al.* 1982; Goodwin *et al.* 1987*a*), sleep pattern (Dugovic *et al.* 1989), sexual function (Ahlenius and Larsson 1985), and hormone secretion (Gilbert *et al.* 1988*a,b*). The present discussion will focus on those behavioural models of 5-HT$_{1A}$ receptor function which have been employed to assess the effects of repeated antidepressant treatments on 5-HT$_{1A}$ receptor sensitivity.

Hypothermia

In the rat, administration of a variety of selective 5-HT$_{1A}$ receptor agonists, including 8-OH-DPAT, gepirone, buspirone, and ipsapirone, causes a decrease in core temperature which is sensitive to antagonism by the 5-HT$_{1A}$ receptor antagonist, pindolol (Goodwin *et al.* 1987*a*; Koenig *et al.* 1988). There has been some controversy surrounding the origin of the 5-HT$_{1A}$ receptor-mediated hypothermic response in the rat. Thus, some authors have reported that 8-OH-DPAT-induced hypothermia is unaffected by central 5-HT depletion (Hjorth 1985; Hutson *et al.* 1987) however, in other studies, a marked attenuation of the hypothermic response following parachlorophenylalanine (PCPA) or 5,7-dihydroxytryptamine (5,7-DHT) treatment was noted (Goodwin *et al.*, 1987*a*; Heal *et al.* 1989). In addition, it has been demonstrated that local injection of 8-OH-DPAT into the dorsal raphe nucleus induces a hypothermic response which is sensitive to antagonism by pindolol (Higgins *et al.* 1988; Hillegaart 1991). Thus, present data suggest that the hypothermic response to 5-HT$_{1A}$ receptor agonists in the rat reflects, at least in part, the activation of 5-HT$_{1A}$ autoreceptors.

5-HT$_{1A}$ receptor-mediated behaviour

Administration of 5-HT$_{1A}$ receptor agonists such as 8-OH-DPAT or gepirone produces a characteristic behavioural syndrome comprising forepaw treading, head-weaving, hind-limb abduction, and hyperlocomotion, which is sensitive to antagonism by 5-HT$_{1A}$ receptor antagonists such as pindolol and propranolol (Hjorth *et al.* 1982; Goodwin *et al.* 1987*a*). The 8-OH-DPAT-induced behavioural syndrome has been shown to be maintained following, 5,7-DHT induced lesions of central 5-HT neurones, suggesting that it is mediated by postsynaptic 5-HT$_{1A}$ receptors (Goodwin *et al.* 1987*a*).

Neuroendocrine responses

In the rat, systemic administration of 5-HT$_{1A}$ receptor agonists, including 8-OH-DPAT, buspirone, gepirone, and ipsapirone, increases the plasma concentration of corticotropin (ACTH) and corticosterone (Koenig *et al.* 1987; Gilbert *et al.* 1988*a*; Przegalinski *et al.* 1989*a*). These responses are attenuated by 5-HT$_{1A}$ receptor antagonists such as pindolol and MDL 73005 (Gilbert *et al.* 1988*a*; Gartside *et al.* 1990). Local administration of 8-OH-DPAT into the hypothalamus also increases plasma corticosterone (Haleem *et al.* 1989), suggesting a postsynaptic location for the 5-HT$_{1A}$ receptors which mediate ACTH release. This is further supported by the findings that the ACTH and corticosterone responses to systemic administration of 8-OH-DPAT are not decreased by central 5-HT depletion, and thus do not rely on the integrity of 5-HT neurones (Gilbert *et al.* 1988*b*; Przegalinski *et al.* 1989*b*).

Whilst the 5-HT$_{1A}$ receptor agonist, buspirone reliably increases plasma prolactin (PRL) (Meltzer *et al.* 1982), this effect is not consistently seen with other 5-HT$_{1A}$ receptor agonists. Thus, 8-OH-DPAT has been reported by some authors to provoke a small increase in plasma prolactin (Simonovic *et al.* 1984; Aulakh *et al.* 1988*a*), but other studies have failed to show such an effect (Di Renzo *et al.* 1989; Van de Kar *et al.* 1989). Two other 5-HT$_{1A}$ receptor agonists, gepirone and ipsapirone do not appear to increase PRL release in the rat (Nash and Meltzer 1989). These inconsistencies have led to the suggestion that the effect of buspirone (and possibly 8-OH-DPAT) to increase plasma PRL may result from blockade of pituitary dopamine D$_2$ receptors (Meltzer *et al.* 1982; Nash and Meltzer 1989).

ANTIDEPRESSANT TREATMENT AND 5-HT$_{1A}$ AUTORECEPTORS IN THE RAT

Electrophysiological and biochemical studies

de Montigny and colleagues (de Montigny *et al.* 1983, 1989) have used electrophysiological methods to study the effects of many different antidepressant treatments on the sensitivity of 5-HT$_{1A}$ autoreceptors in the dorsal raphe nucleus. These investigations have shown that on acute administration, selective 5-HT uptake inhibitors (SSRIs) and monoamine oxidase inhibitors (MAOIs) decrease the rate of spontaneous firing of 5-HT neurones, probably secondarily to an increase in synaptic levels of 5-HT (Blier *et al.* 1986; Chaput *et al.* 1986, 1988). However, on repeated treatment the firing rate of 5-HT neurones is found to return to pretreatment levels. This recovery is associated with the desensitization of 5-HT$_{1A}$ autoreceptors as evidenced by a decrease in the effect of 5-HT$_{1A}$ receptor agonists to depress the firing rate of 5-HT neurones (Table 6.1).

TABLE 6.1. *Effect of antidepressant treatments on 5-HT$_{1A}$ receptor function in the rat*

Model	Antidepressant treatment			
	TCA	SSRI	MAOI	Lithium
5-HT$_{1A}$ autoreceptor				
Hypothermia	↓[a]	↓[a]	↓[a]	0[b]
Electrophysiology	0[c]	↓[d]	↓[e]	0[f]
Postsynaptic 5-HT$_{1A}$ receptor				
Electrophysiology	↑[e]	0[g]	↓[e]	?↑[f]
Adenylate cyclase	↓ h0[i,j]	0[j]	↓[j]	↓ h0[i]
Behaviour	↓[a]	↓[a]	↓[a]	↑[b]
ACTH/corticosterone	0[k,l,m]	?	0[k]	0[m]

[a] Goodwin *et al.* 1987.
[b] Goodwin *et al.* 1986.
[c] de Montigny *et al.* 1989.
[d] Chaput *et al.* 1988.
[e] Blier *et al.* 1986.
[f] Blier *et al.* 1987.
[g] Chaput *et al.* 1986.
[h] Newman *et al.* 1990.
[i] Odagaki *et al.* 1991.
[j] Varrault *et al.* 1991.
[k] Aulakh *et al.* 1988.
[l] Gartside and Cowel, unpublished.
[m] Przegalinski *et al.* 1989*b*.

Chronic administration of lithium or tricyclic antidepressants (TCAs) was not found to be associated with desensitization of 5-HT$_{1A}$ autoreceptors (Blier *et al.* 1987) (Table 6.1). It is not immediately clear why this effect should be absent following TCA treatment, since most TCAs are (albeit, nonselective) inhibitors of 5-HT uptake. However, some confirmation of the electrophysiological data has come from an autoradiographic investigation where the density of 5-HT$_{1A}$ receptors in the dorsal raphe nucleus was significantly decreased by chronic treatment with the SSRI, fluoxetine, but not the TCA, amitriptyline (Welner *et al.* 1989).

Hypothermic response

As discussed above, the hypothermic response to 8-OH-DPAT in the rat probably reflects activation of 5-HT$_{1A}$ autoreceptors in the dorsal raphe nucleus. This response is attenuated by a wide range of antidepressants including MAOIs and SSRIs as well as TCAs (Goodwin *et al.* 1987*b*). In some respects, therefore, these findings resemble data from the electrophysiological studies described above; however, the attenuation of the hypothermic response is induced by a wider range of antidepressant treatments. Repeated lithium treatment, however, does not attenuate the hypothermic response to 8-OH-DPAT in the rat (Goodwin *et al.* 1986) (Table 6.1).

ANTIDEPRESSANT TREATMENT AND POSTSYNAPTIC 5-HT$_{1A}$ RECEPTORS IN THE RAT

Electrophysiological and biochemical studies

Ligand-binding studies have failed to show consistent effects of antidepressant treatments on postsynaptic 5-HT$_{1A}$ receptor density. Thus MAOIs and SSRIs are reported to be without effect on [^{3}H]-8-OH-DPAT binding in hippocampus (Frazer and Hensler 1990). In general, TCAs are also reported not to alter hippocampal 5-HT$_{1A}$ receptor density (Newman *et al.* 1990; Odagaki *et al.* 1991), although one study has shown a decrease following imipramine (Mizuta and Segawa 1988), and another, an increase after amitriptyline (Welner *et al.* 1989).

Activation of postsynaptic 5-HT$_{1A}$ receptors in rat hippocampus induces membrane hyperpolarization and a decrease in the firing rate of pyramidal cells. This electrophysiological response is reported to be unaffected by SSRIs (de Montigny *et al.* 1989), and to be slightly attenuated by treatment with the MAOI clorgyline (Blier *et al.* 1987). Conversely, TCAs on repeated administration have been reported to enhance the inhibitory response of hippocampal pyramidal cells, suggesting increased responsiveness of postsynaptic 5-HT$_{1A}$ receptors (de Montigny *et al.* 1989) (Table 6.1.).

The 5-HT$_{1A}$ receptor-mediated inhibition of forskolin-stimulated adenylate cyclase in hippocampus has been reported to be decreased by MAOIs (Varrault *et al.* 1991) but the effect of TCAs on this response is more variable (Newman *et al.* 1990; Odagaki *et al.* 1991; Varrault *et al.* 1991) (Table 6.1). Whilst these findings appear discrepant with the electrophysiological data, it is possible that the 5-HT$_{1A}$ receptor-mediated membrane hyperpolarization measured in electrophysiological studies may not be linked to modulation of adenylate cyclase activity.

The effects of lithium on postsynaptic 5-HT$_{1A}$ receptors are similarly inconsistent. Thus, in ligand-binding studies of hippocampal 5-HT$_{1A}$ receptor density, lithium has been reported to cause a decrease in [^{3}H]-8-OH-DPAT binding (Odagaki *et al.* 1990) or to be without significant effect

(Newman *et al.* 1990). Interestingly, Newman and Lerer (1988) reported an attenuation of the 5-HT$_{1A}$ receptor-mediated inhibition of forskolin-stimulated adenylate cyclase in hippocampal membranes, whilst Odagaki *et al.* (1991) found this measure of postsynaptic 5-HT$_{1A}$ receptor sensitivity to be unchanged. Blier *et al.* (1987) found no effect of lithium treatment on the sensitivity of hippocampal pyramidal cells to 5-HT agonists but speculated that their data were consistent with an enhanced sensitivity of some subsets of 5-HT$_{1A}$ receptors, located elsewhere in the brain (Table 6.1).

Behavioural and neuroendocrine responses

Repeated treatment of rats with a TCA, the SSRI, zimelidine, and an MAOI attenuated the behavioural syndrome produced by 8-OH-DPAT in the rat (Goodwin *et al.* 1987*b*). In contrast, 8-OH-DPAT-induced behaviours were strikingly enhanced by lithium treatment (Goodwin *et al.* 1986) (Table 6.1).

Contrary to the effects of antidepressants in the behavioural model of postsynaptic 5-HT$_{1A}$ receptor sensitivity, neuroendocrine studies have shown that chronic treatment with TCAs, MAOIs, and lithium does not influence the ACTH/corticosterone response to 5-HT$_{1A}$ receptor stimulation (Aulakh *et al.* 1988*b*; Przegalinski *et al.* 1989*b*; Gartside *et al.* 1992) (Table 6.1).

ANTIDEPRESSANT TREATMENTS AND 5-HT$_{1A}$ NEUROTRANSMISSION IN THE RAT

As discussed above, there is evidence that repeated administration of a number of antidepressant treatments decreases the sensitivity of 5-HT$_{1A}$ autoreceptors in the raphe nuclei. With drugs such as TCAs, 5-HT uptake inhibitors and MAOIs, whose primary pharmacological action is to increase the synaptic availability of 5-HT, a decrease in the inhibitory action of 5-HT on its own release, might then be expected to lead to a sustained increase in the synaptic availability of 5-HT.

The effects of antidepressant drugs on the sensitivity of postsynaptic 5-HT$_{1A}$ receptors are inconsistent and appear to depend on the experimental model employed and the brain region studied. However, taking models which assess 5-HT$_{1A}$ receptor function in hippocampus or hypothalamus, it appears that TCAs and 5-HT uptake inhibitors either increase or do not change postsynaptic 5-HT$_{1A}$ receptor sensitivity. This action at postsynaptic 5-HT$_{1A}$ receptors, taken together with the effect of TCAs and 5-HT uptake inhibitors to increase synaptic 5-HT levels, would be expected to lead to a net enhancement of neurotransmission at postsynaptic 5-HT$_{1A}$ receptors in hippocampus and hypothalamus.

Whilst postsynaptic 5-HT$_{1A}$ receptor-mediated electrophysiological responses in the hippocampus are slightly reduced by chronic MAOI treatment, Blier *et al.* (1986) have shown, using stimulation of ascending 5-HT pathways, that overall postsynaptic 5-HT$_{1A}$ receptor-mediated neurotransmission in the hippocampus is nevertheless enhanced by MAOIs. Repeated lithium treatment similarly enhanced overall 5-HT$_{1A}$ receptor-mediated neurotransmission in this model. From this it seems likely that the facilitation of presynaptic 5-HT function produced by lithium and MAOIs is sufficient to outweigh any small accompanying decrease in postsynaptic 5-HT$_{1A}$ receptor sensitivity.

Overall, therefore, evidence from animal studies indicates that anti-depressant treatments enhance neurotransmission at postsynaptic 5-HT$_{1A}$ synapses though the way in which this effect is produced varies from treatment to treatment. The data suggests that the most consistent effects of antidepressant administration are expressed on presynaptic 5-HT neurones where a combination of acute pharmacological effects and evolving adaptive changes in 5-HT$_{1A}$ autoreceptor sensitivity leads to a sustained increase in the levels of 5-HT in the synaptic cleft. The effects of anti-depressant treatments on postsynaptic 5-HT$_{1A}$ receptors are less consistent but, in general, it appears that any reductions in postsynaptic receptor sensitivity that may occur, do not offset the increase in 5-HT neuro-transmission produced by the facilitation of presynaptic 5-HT neuronal function.

MODELS OF 5-HT$_{1A}$ RECEPTOR FUNCTION IN HUMANS

Neuroendocrine responses

Effect of L-tryptophan

Administration of the 5-HT precursor, L-tryptophan (LTP), to human subjects produces a reliable and dose-related increase in plasma PRL concentration (Cowen *et al.* 1985). This response is enhanced by pretreatment with clomipramine (a TCA which selectively inhibits 5-HT uptake) (Anderson and Cowen 1986), suggesting that it is mediated by brain 5-HT pathways. We have subsequently conducted a series of studies with selective 5-HT receptor antagonists to determine the subtype of postsynaptic 5-HT receptor involved in the PRL response to LTP.

We found that LTP-induced PRL release was not antagonized by pretreatment with ritanserin (a 5-HT$_{2/1C}$ receptor antagonist) (Charig *et al.* 1986) or BRL 43694 (a 5-HT$_3$ receptor antagonist) (Anderson *et al.* 1988). The increase in plasma PRL concentration was, however, significantly inhibited by pindolol (5-HT$_{1A}$ receptor antagonist) (Smith *et al.* 1991).

These findings suggest that the PRL response to LTP is mediated via postsynaptic 5-HT_{1A} receptors though further studies need to be carried out with more selective 5-HT_{1A} receptor antagonists.

Effect of 5-HT₁ₐ receptor agonists

A number of selective 5-HT_{1A} receptor agonists are now available for human use. As in animal studies all these drugs produce an increase in plasma ACTH and cortisol and, in addition, a consistent stimulation of plasma GH is seen (Lesch *et al.* 1989; Cowen *et al.* 1990; Lesch 1991). Effects of these drugs on plasma PRL are more variable with buspirone, in doses of 15 mg or greater, reliably increasing PRL concentrations while the effect of gepirone on this response is less consistent (Anderson *et al.* 1990; Cowen *et al.* 1990). Ipsapirone does not alter PRL levels in humans (Lesch *et al.* 1989). Preliminary studies with pindolol and the 5-HT_{1A} receptor antagonist, MDL 73005, suggest that the ACTH and GH responses to buspirone and ipsapirone are mediated by postsynaptic 5-HT_{1A} receptors (Lesch *et al.* 1990a; Boyce *et al.* 1991; Anderson and Cowen 1992). The increase in plasma PRL produced by buspirone may be partly mediated by 5-HT_{1A} receptors but, as in animal studies, a large component of this response may result from dopamine receptor blockade (Anderson and Cowen 1992).

Hypothermic responses

As in animal studies, administration of 5-HT_{1A} receptor agonists produces a decrease in body temperature which can be detected as a lowering of oral temperature (Cowen *et al.* 1991). In the case of both ipsapirone and buspirone this effect can be attenuated by pindolol (Lesch *et al.* 1990b; Anderson and Cowen 1992) suggesting that it is mediated by activation of 5-HT_{1A} receptors. By analogy with animal studies these receptors would be expected to be 5-HT_{1A} autoreceptors located on raphe cell bodies.

EFFECTS OF ANTIDEPRESSANT TREATMENT ON 5-HT₁ₐ NEUROTRANSMISSION IN HUMANS

Neuroendocrine responses

Studies with L-tryptophan

The PRL response to LTP can be taken as an indication of overall neuro-transmission through postsynaptic 5-HT_{1A} synapses (see above). There is evidence from a number of studies in both depressed patients and healthy subjects that this response is enhanced by treatment with TCAs, MAOIs,

TABLE 6.2. *Effect of antidepressant treatments on 5-HT$_{1A}$ receptor function in humans*

	Antidepressant treatment			
Model	TCA	SSRI	MAOI	Lithium
5-HT$_{1A}$ autoreceptor Hypothermia	↓ [a]	↓ [b]	?	0 [c]
Postsynaptic 5-HT$_{1A}$ receptor Endocrine	0 [d]	↓ [b]	?	0 [c]
Overall 5-HT$_{1A}$ neurotransmission PRL response to LTP	↑ [e,f]	↑ [g]	↑ [h]	↑ [i,j]

[a] Lesch *et al.* 1990*d*.
[b] Lesch *et al.* 1991.
[c] Walsh *et al.* 1991.
[d] Lesch 1991.
[e] Charney *et al.* 1984.
[f] Cowen *et al.* 1986.
[g] Price *et al.* 1989*a*.
[h] Price *et al.* 1987.
[i] Glue *et al.* 1986.

and SSRIs (Charney *et al.* 1984; Price *et al.* 1985, 1989*a*; Cowen *et al.* 1986). No enhancement, however, is seen following repeated administration of trazodone (Price *et al.* 1988) or mianserin (Cowen 1988) (Table 6.2).

Unlike the former compounds, mianserin and trazodone lack significant 5-HT uptake or MAO-inhibiting properties and indeed the extent of the enhancement of the PRL response to LTP seen with different antidepressants seems to correlate with their ability to inhibit 5-HT uptake (Fig. 6.2). Whether this is the sole explanation for their effects, however, requires further study. It is possible, for example, that alterations in 5-HT$_{1A}$ receptor sensitivity may also be involved (see below).

It is of interest that lithium also increases the PRL response to LTP in both healthy subjects and depressed patients (Glue *et al.* 1986; Price *et al.* 1989*b*). In depressed patients taking TCA treatment, the addition of lithium produces a further increase in the PRL response to LTP (Cowen *et al.* 1991). This finding provides support for the hypothesis of de Montigny *et al.* (1983) that the therapeutic effect of lithium–TCA treatment combination in resistant depression may be mediated via a synergistic effect of both compounds to facilitate 5-HT$_{1A}$ neurotransmission (Table 6.2).

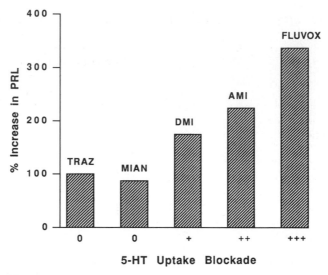

FIG. 6.2. The effect of antidepressant drugs with varying abilities to inhibit 5-HT uptake on the prolactin (PRL) response to L-tryptophan (LTP) in depressed patients. The % increase refers to the change in PRL response to LTP on drug treatment compared to the response before treatment began.

Studies with $5\text{-}HT_{1A}$ receptor agonists

There are, at present, very few studies on the effects of antidepressant treatment on endocrine responses to selective $5\text{-}HT_{1A}$ receptor agonists. Lesch (1991) reported that the ACTH and cortisol responses to ipsapirone were unchanged following amitriptyline treatment in depressed patients, but in view of the pretreatment blunting in ipsapirone-induced ACTH and cortisol release described in depressed subjects (Lesch *et al.* 1990c) this finding is hard to interpret. Lesch *et al.* (1991) have also reported that the ACTH and cortisol responses to ipsapirone are blunted by fluoxetine treatment in patients with obsessive–compulsive disorder (OCD) (Table 6.2).

We found that the endocrine responses to gepirone were not altered by lithium treatment in healthy volunteers (Walsh *et al.* 1991), which suggests that the increase in the PRL response to LTP is not due to an increase in sensitivity of postsynaptic $5\text{-}HT_{1A}$ receptors. Clearly, further studies of this sort need to be carried out to characterize the effect of a range of antidepressant treatments on postsynaptic $5\text{-}HT_{1A}$ receptor responsiveness (Table 6.2).

Hypothermic responses

Lesch *et al.* (1990d) have reported that the hypothermic response to ipsapirone in depressed patients is blunted by treatment with amitriptyline. We have found a similar attenuation of the hypothermic response to buspirone in depressed patients treated for 4–6 weeks with a range of TCAs (Fig. 6.3). Lesch *et al.* (1991) also found that the hypothermic response to ipsapirone was attenuated by fluoxetine in patients with OCD (Table 6.2).

In contrast, treatment of healthy subjects with lithium failed to alter the hypothermic response to gepirone (Walsh *et al.* 1991). Taken together the data suggest that in humans as in rodents, repeated treatment with TCAs and SSRIs may desensitize 5-HT$_{1A}$ autoreceptors.

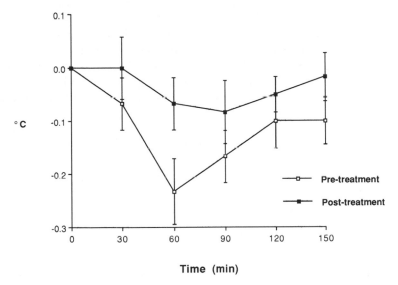

Time (min)

FIG. 6.3. Mean (SEM) change in temperature from baseline produced by buspirone (20–30 mg orally) in six depressed patients. The patients were tested on two occasions (a) drug-free and (b) after 4 weeks treatment with a tricyclic antidepressant (TCA). The peak hypothermic response to buspirone was significantly attenuated by TCA treatment ($p<0.05$, paired *t*-test).

ANTIDEPRESSANT TREATMENTS AND 5-HT$_{1A}$ NEUROTRANSMISSION IN HUMANS

If the PRL response to LTP is taken as a valid measure of 5-HT$_{1A}$ neurotransmission (a conclusion from which some would dissent) (van Praag *et al.* 1986) it appears that a variety of antidepressant treatments may increase 5-HT$_{1A}$ neurotransmission in humans. The mechanism of this effect requires further study but at present it appears as that the ability of a drug to inhibit the uptake of 5-HT (TCAs and 5-HT uptake inhibitors) or to decrease 5-HT metabolism (MAOIs) may be important.

How far changes in 5-HT$_{1A}$ receptor sensitivity may contribute to the enhancement of 5-HT$_{1A}$ neurotransmission produced by some antidepressant treatments in humans remains to be determined. It seems likely that treatment with TCAs and 5-HT uptake inhibitors decreases the sensitivity of 5-HT$_{1A}$ autoreceptors which would be expected to facilitate 5-HT release. It is also possible that changes in postsynaptic 5-HT$_{1A}$ receptor sensitivity could contribute to increases in the PRL response to LTP produced by some antidepressants but as yet there is little human data with which to resolve this issue. In the case of 5-HT uptake inhibitors, however, it appears as though the decrease in postsynaptic 5-HT$_{1A}$ receptor sensitivity revealed by ipsapirone challenge (Lesch *et al.* 1991) is insufficient to outweigh the net effect of these drugs to increase overall 5-HT$_{1A}$ neurotransmission.

It is of interest that lithium treatment also appears to enhance 5-HT$_{1A}$ neurotransmission in humans (Glue *et al.* 1986; Price *et al.* 1989*b*; Cowen *et al.* 1991). From studies with gepirone it seems clear that the ability of lithium, administered alone, to increase 5-HT-mediated PRL release is not mediated by changes in the sensitivity of either 5-HT$_{1A}$ autoreceptors or postsynaptic 5-HT$_{1A}$ receptors (Walsh *et al.* 1991). There is, however, much evidence from animal studies that lithium may facilitate the presynaptic release of 5-HT (Treiser *et al.* 1981; Friedman *et al.* 1988), though the precise biochemical mechanism underlying this effect has not been determined.

ROLE OF 5-HT IN ANTIDEPRESSANT ACTIONS OF ANTIDEPRESSANT DRUGS

The animal and human studies outlined above demonstrate convincingly that many antidepressant drug treatments alter aspects of brain 5-HT function and, in particular, may facilitate neurotransmission at postsynaptic 5-HT$_{1A}$ receptors. These data, however, leave unresolved the question of the relationship of such alterations in 5-HT function to the antidepressant effects of antidepressant treatments. One approach to this problem is to assess the effect of inhibition of brain 5-HT synthesis upon the response to antidepressant drug treatment.

Reversal of antidepressant effects by 5-HT depletion

Use of parachlorophenyalanine (PCPA)

If enhancements in 5-HT neurotransmission are important in mediating the therapeutic effects of antidepressant treatment, it might be supposed that these antidepressant effects could be reversed by measures that inhibit brain 5-HT function. In some early studies Shopsin *et al.* (1975, 1976) found that administration of PCPA to depressed patients who had recently recovered following treatment with either imipramine or tranylcypromine led to a rapid return in depressive symptoms.

Use of tryptophan depletion

These studies have been criticized because of their uncontrolled nature but recent confirmation has been obtained by investigators at Yale, using a dietary manipulation developed by Young *et al.* (1985). When normal subjects are administered a high concentration of oral amino acids (about 100 g) with no tryptophan, plasma tryptophan declines to about 20 per cent of normal values over the next 5 h. From animal studies this would be expected to lead to a rapid reduction in the entry of tryptophan to the brain with a resulting inhibition of brain 5-HT synthesis (Biggio *et al.* 1975).

In normal volunteers this manoeuvre gives rise to a mild but statistically significant lowering of mood (Young *et al.* 1985). However, in patients recently recovered from depressive disorders and maintained on anti-depressant treatments an abrupt and severe recrudescence of depressive symptoms has been reported (Delgado *et al.* 1990). Systematic studies of the effect of tryptophan depletion in relation to particular antidepressant treatments have not yet been published, but preliminary findings suggest that patients receiving MAOIs or 5-HT uptake inhibitors are particularly likely to show a relapse in their depressive symptoms (Delgado *et al.* 1990).

These studies are important in underpinning the role of 5-HT pathways in the effects of antidepressant treatment. They strongly suggest that the actions of some antidepressants are dependent on intact brain 5-HT neurotransmission. Interestingly in the studies of Shopsin *et al.* (1975), depletion of catecholamines with alpha-methyl-para-tyrosine did not lead to a recurrence of depressive symptoms.

CONCLUSIONS

The recent description of multiple 5-HT receptor subtypes has led to a much better understanding of how antidepressant treatments interact with 5-HT neurones. Data from both human and animal studies suggest that many antidepressant treatments, including TCAs, MAOIs, SSRIs, and lithium facilitate neurotransmission at postsynaptic 5-HT$_{1A}$ synapses. A variety of mechanisms may be involved in this action which can be best understood in terms of a combination of acute pharmacological effects together with slowly evolving adaptive changes in both 5-HT and other neurotransmitter systems.

The relationship of this facilitation of brain 5-HT function to therapeutic outcome remains to be clearly established. However, recent studies showing that acute tryptophan depletion causes a rapid relapse in depressed mood in successfully treated depressed patients supports the proposal that 5-HT pathways play a critical role in the initiation and maintenance of the actions of several classes of antidepressant drugs.

ACKNOWLEDGEMENT

The studies of the authors were supported by the Medical Research Council. S.E.G. was an MRC research student.

REFERENCES

Ahlenius, S. and Larsson, K. (1985). Antagonism by lisuride and 8-OHDPAT of 5-HTP-induced prolongation of the performance of male rat sexual behaviour. *European Journal of Pharmacology*, **110**, 379–81.

Anderson, I.M. and Cowen, P.J. (1986). Clomipramine enhances prolactin and growth hormone responses to L-tryptophan. *Psychopharmacology*, **89**, 131–3.

Anderson, I.M. and Cowen, P.J. (1992). Effect of pindolol on endocrine and temperature responses to buspirone in healthy volunteers. *Psychopharmacology*, **106**, 428–32.

Anderson, I.M., Cowen, P.J., and Grahame-Smith, D.G. (1988). The effect of BRL 43694 on the neuroendocrine response to L-tryptophan infusion. *Journal of Psychopharmacology*, **2**, 2.

Anderson, I.M., Cowen, P.J., and Grahame-Smith, D.G. (1990). The effects of gepirone on neuroendocrine function and temperature in humans. *Psychopharmacology*, **100**, 498–503.

Andrade, R. and Nicoll, R.A. (1987). Pharmacologically distinct action of serotonin on single pyramidal neurons of the rat hippocampus recorded *in vitro*. *Journal of Physiology*, **394**, 99–124.

Aulakh, C.S., Wozniak, K.M., Haas, M., Hill, J.L., Zohar, J., and Murphy, D.L. (1988a). Food intake, neuroendocrine and temperature effects of 8-OHDPAT in the rat. *European Journal of Pharmacology*, **146**, 253–9.

Aulakh, C.S., Wozniak, K.M., Hill, J.L., and Murphy, D.L. (1988b). Differential effects of long-term antidepressant treatments on 8-OHDPAT-induced increases in plasma prolactin and corticosterone in rats. *European Journal of Pharmacology*, **156**, 395–400.

Biggio, G., Fadda, F., Fanni, P., Tagliamonte, A., and Gessa, G.L. (1974). Rapid depletion of serum tryptophan, brain tryptophan, serotonin and 5-hydroxy-indoleacetic acid by a tryptophan diet. *Life Sciences*, **14**, 1321–9.

Blier, P., de Montigny, C., and Azzaro, A.J. (1986). Modification of serotonergic and noradrenergic neurotransmission by repeated administration of monoamine oxidase inhibitors: electrophysiological studies in the rat CNS. *Journal of Pharmacology and Experimental Therapeutics*, **227**, 987–94.

Blier, P., de Montigny, C., and Tardif, D. (1987). Short-term lithium treatment enhances responsiveness of postsynaptic 5-HT$_{1A}$ receptors without altering 5-HT autoreceptor sensitivity: An electrophysiological study in the rat brain. *Synapse*, **1**, 225–32.

Bockaert, J., Dumuis, A., Bouhelal, R., Sebben, M., and Cory, R.N. (1987). Piperazine derivatives including the putative anxiolytic drugs, buspirone and ipsapirone, are agonists at 5-HT$_{1A}$ receptors negatively coupled with adenylate cyclase in hippocampal neurons. *Naunyn-Schmeideberg's Archives of Pharmacology*, **335**, 588–92.

Boyce, M.J., Hinze, C., Haegele, K.D., Green, D., and Cowen, P.J. (1991). Initial studies in man to characterise MDL 73,005EF, a novel 5-HT$_{1A}$ receptor ligand and anxiolytic. In *Serotonin: molecular biology, receptors and functional effects* (ed. J.R. Fozard and P.R. Saxena), pp. 472–82. Birkhauser, Basel.

Chaput, Y., de Montigny, C., and Blier, P. (1986). Effects of a selective 5-HT reuptake blocker, citalopram, on the sensitivity of 5-HT autoreceptors: electrophysiological studies in the rat. *Naunyn-Schmeideberg's Archives of Pharmacology*, **333**, 342–5.

Chaput, Y., Blier, P., and de Montigny, C. (1988). Acute and long-term effects of antidepressant 5-HT reuptake blockers on the efficacy of 5-HT neurotransmission: electrophysiological studies in the rat CNS. *Advances in Biological Psychiatry*, **17**, 1–17.

Charig, E.M., Anderson, I.M., Robinson, J.M., Nutt, D.J., and Cowen, P.J. (1986). L-tryptophan and prolactin release: evidence for interaction between 5-HT$_1$ and 5-HT$_2$ receptors. *Human Psychopharmacology*, **1**, 93–7.

Charney, D.S., Heninger, G.R., and Sternberg, D.E. (1984). Serotonin function and the mechanism of action of antidepressant treatment: effects of amitriptyline and desipramine. *Archives of General Psychiatry*, **41**, 359–65.

Cowen, P.J. (1988). Prolactin response to tryptophan during mianserin treatment. *American Journal of Psychiatry*, **145**, 740–1.

Cowen, P.J., Gadhvi, H., Gosden, B., and Kolakowska, T. (1985). Responses of prolactin and growth hormone to L-tryptophan infusion: effects in normal subjects and in schizophrenic patients receiving neuroleptics. *Psychopharmacology*, **86**, 164–9.

Cowen, P.J., Geaney, D.P., Schachter, M., Green, A.R., and Elliott, J.M. (1986). Desipramine treatment in normal subjects: effects on neuroendocrine responses to tryptophan and on platelet serotonin-related receptors. *Archives of General Psychiatry*, **43**, 61–7.

Cowen, P.J., Anderson, I.M., and Grahame-Smith, D.G. (1990). Neuroendocrine effects of azapirones. *Journal of Clinical Psychopharmacology*, **10**, 21S-5S.

Cowen, P.J., McCance, S.L., Ware, C.J., Cohen, P.R., Chalmers, J.S., and Julier D.L. (1991). Lithium in tricyclic resistant depression: correlation of increased brain 5-HT function with clinical outcome. *British Journal of Psychiatry*, **159**, 341–6.

Delgado, P.L., Charney, D.S., Price, L.H., Aghajanian, D.K., Landis, H., and Heninger, G.R. (1990). Serotonin function and the mechanism of antidepressant action: reversal of antidepressant-induced remission by rapid depletion of plasma tryptophan. *Archives of General Psychiatry*, **47**, 411–18.

de Montigny, C., Cournoyer, G., Morissette, R., Langlois, R., and Caille, G. (1983). Lithium carbonate addition in tricyclic antidepressant-resistant unipolar depression. *Archives of General Psychiatry*, **40**, 1327–34.

de Montigny, C., Chaput, Y., and Blier, P. (1989). Long-term tricyclic and electro-convulsive treatment increases responsiveness of dorsal hippocampus 5-HT$_{1A}$ receptors: an electrophysiological study in the rat. *Society for Neuroscience Abstracts*, **15**, 854.

De Vivio, M. and Maayani, S. (1986). Characterization of the 5-hydroxy-tryptamine$_{1A}$-receptor-mediated inhibition of forskolin-stimulated adenylate cyclase activity in guinea pig and rat hippocampal membranes. *Journal of Pharmacology Experimental Therapeutics*, **238**, 248–53.

Di Renzo, G., Amoroso, S., Taglialatela, M., Canzoniero, L., Basile, V., Fatatis, A. *et al.* (1989). Pharmacological characterization of serotonin receptors involved in the control of prolactin secretion. *European Journal of Pharmacology*, **162**, 371–3.

Dugovic, C., Wauquier, A., Leysen, J.E., Marrannes, R., and Janssen, P.A.J. (1989). Functional role of 5-HT$_2$ receptors in the regulation of sleep and wakefulness in the rat. *Psychopharmacology*, **97**, 436–42.

Frazer, A. and Hensler, J.G. (1990). 5-HT$_{1A}$ receptors and 5-HT$_{1A}$-mediated responses: effect of treatments that modify serotonergic neurotransmission. *Annals of the New York Academy of Sciences*, **600**, 460–75.

Friedman, E. and Wang, H.Y. (1988). Effect of chronic lithium treatment on 5-hydroxytryptamine autoreceptors and release of 5-[^{3}H]hydroxytryptamine from rat brain cortical, hippocampal and hypothalamic slices. *Journal of Neurochemistry*, **50**, 195–201.

Gartside, S.E., Cowen, P.J., and Hjorth, S. (1990). Effects of MDL 73005EF on central pre- and postsynaptic 5-HT$_{1A}$ receptor function in the rat *in vivo*. *European Journal of Pharmacology*, **191**, 391–400.

Gartside, S.E., Ellis, P.M., Sharp, T., and Cowen, P.J. (1992). Selective 5-HT$_{1A}$ and 5-HT$_2$ receptor-mediated adrenocorticotropin release in the rat: effect of repeated antidepressant treatments. *European Journal of Pharmacology*, **221**, 27–33.

Gilbert, F., Brazell, C., Tricklebank, M.D., and Stahl, S.M. (1988a). Relationship of increased food intake and plasma ACTH levels to 5-HT$_{1A}$ receptor activation in rats. *Psychoneuroendocrinology*, **13**, 471–8.

Gilbert, F., Brazell, C., Tricklebank, M.D., and Stahl, S.M. (1988b). Activation of the 5-HT$_{1A}$ receptor subtype increases rat plasma ACTH concentration. *European Journal of Pharmacology*, **147**, 431–9.

Glue, P.W., Cowen, P.J., Nutt, D.J., Kolakowska, T., and Grahame-Smith, D.G. (1986). The effect of lithium on 5-HT-mediated neuroendocrine responses and platelet 5-HT receptors. *Psychopharmacology*, **90**, 398–402.

Goodwin, G.M., De Souza, R.J., Wood, A.J., and Green, A.R. (1986). The enhancement by lithium of the 5-HT$_{1A}$-mediated serotonin syndrome in the rat: evidence for a post-synaptic mechanism. *Psychopharmacology*, **90**, 488–93.

Goodwin, G.M., De Souza, R.J., Green, A.R., and Heal, D.J. (1987a). The pharmacology of the behavioural and hypothermic responses of rats to 8-hydroxy-2-(di-*n*-propylamino) tetralin (8-OH-DPAT). *Psychopharmacology* **91**, 506–11.

Goodwin, G.M., De Souza, R.J., and Green, A.R. (1987*b*). Attenuation by electroconvulsive shock and antidepressant drugs of the 5-HT$_{1A}$ receptor-mediated hypothermia and serotonin syndrome produced by 8-OH-DPAT in the rat. *Psychopharmacology*, **91**, 500–5.

Haleem, D.J., Kennett, G.A., Whitton, P.S., and Curzon, G. (1989) 8-OH-DPAT increases corticosterone but not other 5-HT$_{1A}$ receptor-dependent responses more in females. *European Journal of Pharmacology*, **164**, 435–43.

Hall, M.D., El Mestikawy, S., Emerit, M.B., Pichat, L., Hamon, M., and Gozlan, H. (1985). [^{3}H]8-Hydroxy-2-(di-*n*-propylamino)tetralin binding to pre- and post-synaptic 5-hydroxytryptamine sites in various regions of the brain. *Journal of Neurochemistry*, **44**, 1685–96.

Heal, D.J., Prow, M.R., Martin, K.F., and Buckett, W.R. (1989). Effects of lesioning central 5-HT and noradrenaline containing neurones on 8-OH-DPAT hypo-thermia, mydriasis and hypoactivity. *British Journal of Pharmacology*, **97**, 406P.

Higgins, G.A., Bradbury, A.J., Jones, B.J., and Oakley, N.R. (1988). Behavioural and biochemical consequences following activation of 5-HT$_{1A}$-like and GABA receptors in the dorsal raphe nucleus of the rat. *Neuropharmacology*, **27**, 993–1001.

Hillegaart, V. (1991). Effects of local application of 5-HT and 8-OHDPAT into the dorsal and median raphe nuclei on core temperature in the rat. *Psychopharmacology*, **103**, 291–6.

Hjorth, S. (1985). Hypothermia in the rat induced by the potent serotoninergic agent 8-OH-DPAT. *Journal of Neural Transmission*, **61**, 131–5.

Hjorth, S., Carlsson, A., Lindberg, P., Sanchez, D., Wilkstrom, H., Arvidsson, L-E. *et al.* (1982). 8-Hydroxy-2-(di-*n*-propylamino) tetralin, 8-OHDPAT, a potent and selective simplified ergot congener with central 5-HT-receptor stimu-lating activity. *Journal of Neural Transmission*, **55**, 169–88.

Hoyer, D., Pazos, A., Probst, A., and Palacios, J.M. (1986). Serotonin receptors in the human brain. I. Characterization and autoradiographic localization of 5-HT$_{1A}$ recognition sites. Apparent absence of 5-HT$_{1B}$ recognition sites. *Brain Research*, **376**, 85–96.

Hutson, P., Donohoe, T.P., and Curzon, G. (1987). Hypothermia induced by the putative 5-HT$_{1A}$ agonists LY 165163 and 8-OHDPAT is not prevented by 5-HT depletion. *European Journal of Pharmacology*, **143**, 221–8.

Innis, R.B., Nestler, E.J., and Aghajanian, G.K. (1988). Evidence for G protein mediation of serotonin- and GABA$_B$-induced hyperpolarization of rat dorsal raphe neurons. *Brain Research*, **459**, 27–36.

Jenkins, S.W., Robinson, D.S., Fabre, L.E., Andary, J.J., Messina, M., and Reich, L.A. (1990). Gepirone in the treatment of major depression. *Journal of Clinical Psychopharmacology*, **10**, 77S–86S.

Knapp, J.E. (1985). Clinical profile of buspirone. *British Journal of Clinical Practice*, **39**, 95–105.

Koenig, J.I., Gudelsky, G.A., and Meltzer, H.Y. (1987). Stimulation of corti-costerone and β-endorphin secretion in the rat by selective 5-HT receptor subtype activation. *European Journal of Pharmacology*, **137**, 1–8.

Koenig, J.I., Meltzer, H.Y., and Gudelsky, G.A. (1988). 5-hydroxytryptamine-$_{1A}$ receptor mediated effects of buspirone, gepirone and ipsapirone. *Pharmacology, Biochemistry and Behaviour*, **29**, 711–15.

Lesch, K.P. (1991). 5-HT$_{1A}$ receptor responsivity in anxiety disorders and depression. *Progress in Neuro-Psychopharmacology and Biological Psychiatry*, **15**, 723–33.

Lesch, K.P., Rupprecht, R., Poten, B., Muller, U., Sohnle, K., Fritze, J., and Schulte, H.M. (1989). Endocrine responses to 5-hydroxytryptamine$_{1A}$ receptor activation by ipsapirone in humans. *Biological Psychiatry*, **26**, 203–5.

Lesch, K.P., Sohnle, K., Poten, B., Ruprecht, R., Schoellnhammer, G., and Schulte, H.M. (1990a). Corticotropin and cortisol secretion following central 5-hydroxtryptamine-1A (5-HT$_{1A}$) receptor activation: effects of 5-HT receptor and beta-adrenoceptor antagonists. *Journal of Clinical Endocrinology and Metabolism*, **70**, 670–4.

Lesch, K.P., Poten, B., Sohnle, K., and Schulte, H.M. (1990b) Pharmacology of the hypothermic response to 5-HT$_{1A}$ receptor activation in humans. *European Journal of Clinical Pharmacology*, **39**, 17–19.

Lesch, K.P., Mayer, S., Disselkamp-Tietze, J., Hoh, A., Wiesmann, M., Osterheider, M., and Schulte, H. (1990c) 5-HT$_{1A}$ receptor responsivity in unipolar depression: evaluation of ipsapirone-induced ACTH and cortisol secretion in depressed patients. *Biological Psychiatry*, **28**, 620–8.

Lesch, K.P., Disselkamp-Tietze, J., and Schmidtke, A. (1990d). 5-HT$_{1A}$ receptor function in unipolar depression: effect of chronic amitriptyline treatment. *Journal of Neural Transmission*, **180**, 157–61.

Lesch, K.P., Hoh, A.H., Schulte, H.M., Osterheider, M., and Muller T. (1991). Long-term fluoxetine treatment decreases 5-HT$_{1A}$ receptor responsivity in obsessive-compulsive disorder. *Psychopharmacology*, **105**, 410–20.

Marcinkiewicz, M., Vergé, D., Gozlan, H., Pichat, L., and Hamon, M. (1984). Autoradiographic evidence for the heterogeneity of 5-HT$_{1A}$ sites in the rat brain. *Brain Research*, **291**, 159–63.

Meltzer, H.Y., Simonovic, M., Fang, V.S., and Gudelsky, G.A. (1982). Effect of buspirone on rat plasma prolactin levels and striatal dopamine turnover. *Psychopharmacology*, **78**, 49–53.

Mizuta, T. and Segawa, T. (1988). Chronic effects of imipramine and lithium on postsynaptic 5-HT$_{1A}$ and 5-HT$_{1B}$ sites and on presynaptic 5-HT$_3$ sites in rat brain. *Japanese Journal of Pharmacology*, **47**, 107–13.

Nash, J.F. and Meltzer, H.Y. (1989). Effect of gepirone and ipsapirone on the stimulated and unstimulated secretion of prolactin in the rat. *Journal of Pharmacology and Experimental Therapeutics*, **249**, 236–41.

Newman, M.E. and Lerer, B. (1988). Chronic electroconvulsive shock and desimipramine reduce the degree of inhibition by 5-HT and carbachol of forskolin-stimulated adenylate cyclase in rat hippocampal membranes. *European Journal of Pharmacology*, **148**, 257–60.

Newman, M.E., Drummer, D., and Lerer, B. (1990). Single and combined effects of desimipramine and lithium on serotonergic receptor number and second messenger function in rat brain. *Journal of Pharmacology and Experimental Therapeutics*, **252**, 826–31.

Odagaki, Y., Koyama, T., Matsubara, S., Matsubara, R., and Yamashita, I. (1990). Effects of chronic lithium treatment on serotonin binding sites in rat brain. *Journal of Psychiatry Research*, **24**, 271–7.

Odagaki, Y., Koyama, T., and Yamashita, I. (1991). No alterations in the 5-HT₁ₐ-mediated inhibition of forskolin-stimulated adenylate cyclase activity in the hippocampal membranes from rats chronically treated with lithium or anti-depressants. *Journal of Neural Transmission*, **86**, 85–96.

Pazos, A. and Palacios, J.M. (1985). Quantitative autoradiographic mapping of serotonin receptors in the rat brain. I. Serotonin-1 receptors. *Brain Research*, **346**, 205–30.

Pazos, A., Probst, A., and Palacios, J.M. (1987). Serotonin receptors in the human brain. III. Autoradiographic mapping of serotonin-1 receptors. *Neuroscience*, **21**, 97–122.

Price, L.H., Charney, D.S., and Heninger, G.R. (1985). Effects of tranylcypromine treatment on neuroendocrine, behavioural, and autonomic responses to trypto-phan in depressed patients. *Life Sciences*, **37**, 809–18.

Price, L.H., Charney, D.S., and Heninger, G.R. (1988). Effect of trazodone treatment on serotonergic function in depressed patients. *Psychiatry Research*, **24**, 165–75.

Price, L.H., Charney, D.S., Delgado, P.L., Anderson, G.M., and Heninger, G.R. (1989*a*). Effects of desipramine and fluvoxamine treatment on the prolactin response to tryptophan. *Archives of General Psychiatry*, **46**, 625–31.

Price, L.H., Charney, D.S., Delgado, P.L., and Heninger, G.R. (1989*b*). Lithium treatment and serotonergic function: neuroendocrine and behavioural responses to intravenous tryptophan in affective disorder. *Archives of General Psychiatry*, **46**, 13–19.

Przegalinski, E., Budzisewska, B., Warchol-Kania, A., and Blaszczynska, E. (1989*a*). Stimulation of corticosterone secretion by the selective 5-HT₁ₐ receptor agonist 8-hydroxy-2-(di-*n*-propylamino) tetralin (8-OHDPAT) in the rat. *Pharmacology, Biochemistry and Behaviour*, **33**, 329–34.

Przegalinski, E., Warchol-Kania, A., and Budziszewska, B. (1989*b*). The lack of effect of repeated treatment with antidepressant drugs on the 8-OH-DPAT-induced increase in the serum corticosterone concentration. *Polish Journal of Pharmacology and Pharmacy*, **41**, 63–8.

Sharp, T., Bramwell, S.R., and Grahame-Smith, D.G. (1989*a*). 5-HT₁ agonists reduce 5-hydroxytryptamine release in rat hippocampus *in vivo* as determined by brain microdialysis. *British Journal of Pharmacology*, **96**, 283–90.

Sharp, T., Bramwell, S.R., Hjorth, S., and Grahame-Smith, D.G. (1989*b*). Pharmacological characterization of 8-OH-DPAT-induced inhibition of rat hip-pocampal 5-HT release *in vivo* as measured by microdialysis. *British Journal of Pharmacology*, **98**, 989–97.

Shopkin, B., Gershon, S., Goldstein, J., Friedman, E., and Wilk, S. (1975). Use of synthesis inhibitors in defining a role for biogenic amines during imipramine treat-ment in depressed patients. *Psychopharmacology Communications*, **1**, 239–49.

Shopkin, B., Friedman, E., and Gershon, S. (1976). Parachlorophenylalanine reversal of tranylcypromine effects in depressed patients. *Archives of General Psychiatry*, **33**, 811–19.

Simonovic, M., Gudelsky, G.A., and Meltzer, H.Y. (1984). Effect of 8-hydroxy-2-(di-*n*-propylamino)tetralin on rat prolactin secretion. *Journal of Neural Transmission*, **59**, 143–9.

Smith, C.E., Ware, C.J., and Cowen, P.J. (1991). Pindolol decreases prolactin and growth hormone responses to intravenous L-tryptophan. *Psychopharmacology*, **103**, 140–2.

Sprouse, J.S. and Aghajanian, G.K. (1987). Electrophysiological responses of serotonergic dorsal raphe neurones to 5-HT$_{1A}$ and 5-HT$_{1B}$ agonists. *Synapse*, **1**, 3–9.

Treiser, S.L., Cascio, C.S., O'Donahue, T.L., Thoa, N.B., Jacobwitz, D.M., and Kellar, K.J. (1981). Lithium increases serotonin release and decreases serotonin receptors in hippocampus. *Science*, **213**, 1529–31.

Van de Kar, L.D., Lorens, S.A., Urban, J.H., and Bethea, C.L. (1989). Effect of selective serotonin (5-HT) agonists and 5-HT$_2$ antagonists on prolactin secretion. *Neuropharmacology*, **28**, 299–305.

van Praag, H.M., Lemus, C., and Kahn, R. (1986). The pitfalls of serotonin precursors as challengers in hormonal probes of central serotonin activity. *Psychopharmacology Bulletin*, **22**, 565–70.

Varrault, A., Leveil, V., and Bockaert, J. (1991). 5-HT$_{1A}$-sensitive adenylyl cyclase of rodent hippocampal neurons: effects of antidepressant treatments and chronic stimulation with agonists. *Journal of Pharmacology and Experimental Therapeutics*, **257**, 433–8.

Vergé, D., Daval, G., Patey, A., Gozlan, H., El Mestikawy, S., and Hamon, M. (1985). Presynaptic 5-HT autoreceptors on serotonergic cell bodies and/or dendrites but not terminals are of the 5-HT$_{1A}$ subtype. *European Journal of Pharmacology*, **113**, 463–4.

Walsh, A.E., Ware, C.J., and Cowen, P.J. (1991). Lithium and 5-HT$_{1A}$ receptor sensitivity: a neuroendocrine study in healthy volunteers. *Psychopharmacology*, **105**, 568–72.

Weissmann-Nanoupolus, D., Mach, E., Magre, J., Demassey, Y., and Pujol, J.F. (1985). Evidence for the localization of 5-HT$_{1A}$ binding sites on serotonin containing neurons in the raphe dorsalis and raphe centrales nuclei of the rat brain. *Neurochemistry*, **7**, 1061–72.

Welner, S.A., de Montigny, C., Desroches, J., Desjardins, P., and Surayani-Cadotte, B.E. (1989). Autoradiographic quantification of serotonin$_{1A}$ receptors in rat brain following antidepressant drug treatment. *Synapse*, **4**, 347–52.

Young, S.N., Smith, S.E., Pihl, R., and Ervin, F.R. (1985). Tryptophan depletion causes a rapid lowering of mood in normal males. *Psychopharmacology*, **87**, 173–7.

7

Three distinct roles of 5-HT in anxiety, panic, and depression

J.F.W. DEAKIN

INTRODUCTION

It will be evident from other chapters in this book that 5-HT is implicated in an almost absurd variety of disorders: anorexia, Alzheimer's disease, autism, alcoholism, aggression, attention deficit disorder, arson, to mention only those that begin with 'A'. How it is possible that 5-HT is involved in so many conditions? The key to understanding the role of 5-HT may be to step back from the clinical evidence and to create a general framework or a deeper structure based on what is known about the normal psychological functions and processes which are mediated by 5-HT. In this way some understanding of how symptoms arise from dysfunction may emerge. This chapter concentrates on the role of 5-HT in panic, anxiety, and depression and will touch upon some of the further shores of 5-HT-related disorders.

If there is an insight in this chapter, it is that we can regard panic, anxiety, and depression as disturbances in brain mechanisms of defence. These mechanisms serve to reduce the impact or prevent the occurrence of aversive events (Deakin and Graeff 1991); to use a looser terminology, mechanisms which serve to minimize the impact of adversity, life events, and stress. It will be argued that 5-HT has a central role in these defence mechanisms. However, 5-HT systems are highly differentiated with distinct anatomical pathways and receptor subtypes. Thinking of the 5-HT system as a unitary system does not resolve the many contradictions about the behavioural and clinical evidence. For example, there is evidence that 5-HT facilitates anxiety and that 5-HT function is reduced in depression. Since anxiety and depression typically coexist (Goldberg *et al.* 1987) 5-HT cannot have a unitary role in anxiety and depression. Clearly, it is necessary to embrace the complexity of the system. It is also necessary to recognize there are different types of aversion. To consider the 5-HT system as a 'punishment' system is no longer adequate: there are different kinds of aversion which evoke different coping or defence mechanisms. When these defence mechanisms go wrong different psychopathologies result.

OVERVIEW OF AVERSIVE STIMULI AND DEFENSIVE RESPONSES

Aversive stimuli can be divided into short term and long term, or, acute and chronic—they evoke different coping mechanisms (Fig. 7.1). Acute aversive events can evoke different behavioural responses depending on the imminence of aversion. Proximal aversive stimuli are unconditioned, noxious stimuli such as pain and asphyxia. These unconditioned stimuli (UCS) powerfully evoke the escape or flight reflex, and, if another animal is present, fighting behaviour. These behaviours terminate contact with the stimulus. The reflex is mediated by the brain aversion system (BAS)—a system of neurones running from amygdala (rostral) through hypothalamus and periaqueductal grey matter (caudal). This may be regarded as the archetypal, prototypic mechanism of aversion. Panic can be seen as a disturbance in the functioning of the brain aversion system (Graeff 1990).

Distal aversive stimuli warn of noxious unconditioned stimuli; they indi-

Acute		Chronic
Proximal	Distal	Prox/distal
(UCS)	(UCS/CS)	
asphyxia, pain	sight, sound	pain/threat
	smell	loss
↓	↓	↓
Reflexes	Preparation	Adaptation
escape - fight	avoidance	tolerance
panic	fear	resilience
BAS	DRN	MRN

FIG. 7.1. Aversive stimuli and coping mechanisms (See text for abbreviations).

cate risk of noxious stimulation in the future unless some action is taken. Distal aversive stimuli are threats detected at a distance by distance receptors—sight, sound, and smell. They evoke anticipatory anxiety. Some distal stimuli such as the sight, the sound, the smell of a predator, are hard-wired to evoke fear responses in many species. Other distal stimuli acquire aversiveness by having been associated with unconditioned stimuli; they become conditioned fear stimuli (CS). In both cases, distal aversive stimuli evoke the central motive state of anticipatory anxiety which motivates avoidance behaviour to reduce the chances of the unconditioned stimulus occurring. Much evidence indicates that conditioned fear stimuli elicit anticipatory or preparatory defensive behaviour through the rostral parts of the brain aversion system, particularly the amygdala (Blanchard and Blanchard 1972; Shibata *et al.* 1982; Davis 1989). It will be argued that the central motive state of anxiety is modulated by one of the major forebrain 5-HT systems—that arising from the dorsal raphe nucleus. Disturbances of this system result in excessive or morbid anxiety as in generalized anxiety disorder.

When the acute protective mechanisms fail to terminate the aversive event, aversion may become chronic. Chronic aversive events, like acute, evoke adaptive behaviours (Fig. 7.1). Mechanisms come into play to disconnect, reduce or attenuate the acute response to aversion when the aversion is repeated. This is a form of adaptation or tolerance which results in resilience in the face of adversity. It will be argued that this defence is mediated by median raphe 5-HT projections in the forebrain. When the resilience system breaks down then symptoms of learned helplessness and depression result.

PROXIMAL AVERSION

Unconditioned aversive stimuli evoke the defensive triad of autonomic activation, analgesia and vigorous fight/flight behaviour. This behavioural triad can be fully evoked by electrical stimulation of a system of neurones known as the brain aversion system. It comprises the periaqueductal grey matter (PAG), the periventricular grey matter around the third ventricle and its most rostral extension is the amygdala (Fernandez and Hunsperger 1962). These structures have been stimulated in conscious humans during neurosurgical operations and this induces autonomic activation and intense feelings of anxiety including a strong desire to escape (Nashold *et al.* 1969). It is not a particularly original suggestion to postulate that a panic attack results from spontaneous activation of the brain aversion system in the absence of a noxious stimulus (Graeff 1990). This would occur through some neuronal instability or an epileptiform discharge in the structure.

Patients often remember their first panic attack because without warning there is a sudden and full-blown escape response in the bewildering absence of any trigger.

DISTAL AVERSION SYSTEM

Distal aversive stimuli are stimuli which warn of impending noxious stimulation. They are detected by distance receptors and evoke components of the brain aversion system reflex including autonomic activation and analgesia. As Bolles and Fanselow (1980) have said—fear inhibits pain. However, in contrast to proximal aversion there is inhibition of fight/flight behaviour. Anticipatory anxiety evokes behavioural inhibition and freezing in animals. The Bolles and Fanselow dictum may be extended: fear inhibits escape (Deakin and Graeff 1991). This allows the expression of learned avoidance behaviours to remove the organism from danger. There is evidence that both these effects of anticipatory anxiety (inhibition of escape, and avoidance) are facilitated by projections of the dorsal raphe nucleus modulating the brain aversion system and they are discussed below.

FEAR INHIBITS ESCAPE; ANXIETY INHIBITS PANIC

Distal aversive stimuli

The dorsal raphe nucleus is embedded in the ventral part of the periaqueductal grey matter. The PAG receives an innervation both from the dorsal raphe nucleus and the median. Kiser *et al.* (1980) showed that stimulation of the PAG evoked powerful escape behaviour which could be blocked by concurrent stimulation of the dorsal raphe nucleus. Since then, many studies unanimously report that stimulation-induced escape is inhibited by drugs which increase 5-HT neurotransmission (Schutz *et al.* 1985; Audi *et al.* 1988; Jenck *et al.* 1989). However, the receptor subtype involved in restraint of the escape mechanism is not clear. Some evidence suggests a role of $5HT_{1A}$ receptors in restraining the escape mechanism but other lines of evidence suggest $5\text{-}HT_2$ receptors may be involved.

If fear inhibits the fight/flight response, the human corollary is that anticipatory anxiety inhibits panic. This seems a strange suggestion. However, it may make sense of otherwise paradoxical clinical phenomena. Some patients with panic disorder dislike doing relaxation exercises because they precipitate panic attacks (Adler *et al.* 1987). Patients with panic disorder prefer to live in a state of activity and tension perhaps because this restrains spontaneous activation of the brain aversion system and thus panic. Klein and Klein (1989) described how, in the evolution of agoraphobia, panic attacks gradually reduce in frequency as anticipatory anxiety increases. This may be the result of the proposed physiological antagonism between anticipatory anxiety and panic.

If 5-HT restrains the fight/flight mechanism then drugs which enhance 5-HT function should reduce the symptoms of panic and drugs which block 5-HT should disinhibit spontaneous activation of the brain aversion system. The clinical evidence is examined below.

5-HT inhibits panic

It is well established that antidepressant drugs, particularly those acting on 5-HT reuptake, have antipanic activity. Den Boer and Westenberg (1990), for example, showed that fluvoxamine caused progressive reduction in panic attacks in patients with panic disorder over a period of 8 weeks, compared to patients on placebo. In contrast, a group of patients treated with the $5\text{-HT}_2./_{1C}$ antagonist ritanserin showed no improvement and deteriorated on some ratings.

In a study conducted at Manchester University, 56 patients with the generalized neurotic syndrome were recruited into a trial of ritanserin against placebo. Patients with major depression were excluded but the presence of significant anxiety and/or depressive symptoms was required. At the time the trial was conducted the investigators believed that the neuroses were not divisible and so no more detailed diagnoses were made. However, in view of the idea above that 5-HT might have very different roles in different forms of anxiety, patients were retrospectively allocated DSM-III diagnoses. Figure 7.2 shows that patients with panic diagnoses improved less on ritanserin than they did on placebo. This is compatible with the suggestion that 5-HT restrains an unstable brain aversion system which underlies panic disorders. However, panic attacks had not been specifically rated; it is therefore uncertain whether ritanserin had disinhibited panic or some other aspect of anxiety.

5-HT, anxiety, and primary drive disorders

In addition to restraining unconditioned escape responses mediated by the brain aversion system, anticipatory anxiety restrains other primary drives such as sexual and aggressive behaviour, perhaps by 5-HT projections to the hypothalamus. Impaired 5-HT function might thus result in sexual and aggressive disorders. Antisocial personality disorder is characterized by impulsive behaviour and the inability to restrain delayed gratification. These traits are normally restrained by fear of consequences, according to

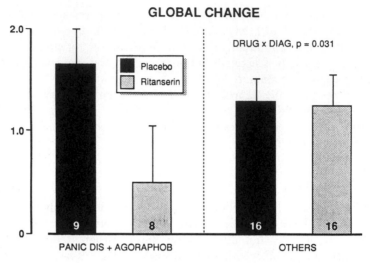

FIG. 7.2. Reanalysis of the Manchester DSM-III panic versus nonpanic diagnoses. The diagnoses were made retrospectively from the case notes. Panickers on ritanserin improved less than those on placebo.

the present theory through 5-HT projections. This would be in keeping with the impulsivity – low 5-HT connection (Linnoila and Virkkunen 1991). The distractibility and euphoria of mania are almost by definition a reflection of excessive incentive motivation. Anxiety restrains incentive, according to the present theory, through 5-HT systems.

It may be that in some cases 'primary drive disorders' (Table 7.1) arise through a loss of restraint by anxiety and the dorsal raphe 5-HT mechanism. However, at least in the case of panic and other phasic disorders of sexual and aggressive behaviour, the primary disturbance is perhaps more plausibly seen as a neuronal instability in the drive centres themselves. It may simply be fortunate that there is a physiological restraint of these

TABLE 7.1. *Primary drive disorders.*

Escape	Panic
Sex	Impulsive sexual assault
Aggression	Impulsive aggression
Reward	Psychopathy Mania

Restrained by anxiety – 5-HT – DRN.

mechanisms by 5-HT. This allows the possibility of therapeutic interventions with drugs which enhance 5-HT functioning.

The dorsal raphe nucleus and anticipatory anxiety

We have discussed one function of anticipatory anxiety which is to restrain the fight/flight mechanism and its human equivalent—panic. Anticipatory anxiety also functions as a drive which motivates avoidance behaviour. This, it will be argued, involves the forebrain terminals of the dorsal raphe system. These terminals show a striking parallelism with the distribution of dopaminergic terminals in frontal cortex, amygdala, and basal ganglia (Azmitia *et al.* 1978). There is evidence that postsynaptic 5-HT$_2$ receptors show a specific association with terminals of the dorsal raphe system (Blue *et al.* 1988). This also appears to be true of the 1D receptor although some of these are presynaptic on 5-HT terminals. $5HT_{1C}$ and $5HT_3$ receptors also have a distribution which is similar to that of terminals of the dorsal raphe system.

Many animal behavioural experiments have shown that interfering with neurotransmission in the dorsal raphe pathway interferes with the expression of behaviours associated with anticipatory anxiety (Hodges *et al.* 1987). The results are compatible with the idea that the dorsal raphe system is activated in anticipatory anxiety. There is little direct evidence of this. Many stressful stimuli activate 5-HT release but recordings of neuronal activity from the dorsal raphe has not shown changes in neuronal firing rate in association with anxiety-related behaviours. An exception is a paper by Walletschek and Raab (1982) who showed striking increases in dorsal raphe firing rate in freely behaving tree shrews when threatened by the approaching hand of the experimenter. Recent results using *in vivo* dialysis to monitor 5-HT release in freely behaving animals show that not only tail pinch and handling but also the more naturalistic anxiogenic stimulus of being placed on the exposed arms of a plus maze, evoke 5-HT release (Kalen *et al.* 1989; Pei *et al.* 1990; Marsden, personal communication). These findings are compatible with the theory that the dorsal raphe 5-HT system may function as a negative incentive system which guides the organism away from danger signals. It may work in opposition to the dopaminergic system where the evidence is now quite compelling that dopamine has the opposite function of homing-in the organism onto reinforcers by tracking incentives in the environment (Fibiger and Phillips 1988).

Morbid anxiety states such as generalized anxiety and anxiety associated with depression involve anticipatory anxiety and may be due to excessive neurotransmission into the dorsal raphe system. This may arise either because there are a large number of threats in the individuals environment or because the system is excessively sensitive to minor threats. If this is true, then drugs which reduce neurotransmission in this pathway should improve anxiety states. Two classes of anxiolytic drugs appear to do this by acting presynaptically on 5-HT neurones themselves. Buspirone, the 1A autoreceptor agonist, causes an acute suppression of the firing of dorsal raphe units. Median raphe units are less sensitive. The fact that the anxiolytic effects of buspirone take time to emerge may reflect the possibility that full antagonism of neurotransmission through dorsal raphe synapses requires the gradual down regulation of 5-HT$_2$ and 5-HT$_3$ receptors that the drug causes after chronic administration. It has been known for many years that benzodiazepines have the ability to suppress firing of dorsal raphe units. It is unlikely, however, that this is entirely responsible for its anxiolytic effect in view of the widespread distribution of benzodiazepine receptors in the amygdala and other anxiety-related structures.

Another approach to the treatment of anxiety would be to block the postsynaptic receptors associated with the dorsal raphe pathway (see above). Indeed, there is a good deal of interest in 5-HT$_2$, 1C, and 3 receptor antagonists as anxiolytics but the clinical evidence is either disappointing or scanty (Ceulemans *et al.* 1985; Critchley and Handley 1986; Costall *et al.* 1990). Clinically, the most effective anxiolytic drugs are undoubtedly antidepressants particularly the older reuptake blockers (Johnstone *et al.* 1980). It has been previously suggested that the shared ability of these drugs either to block or to down regulate 5-HT$_2$/$_{1C}$ receptor function may contribute their anxiolytic activity (Deakin 1988). To this may be added the more recent finding that they also down regulate 5-HT$_3$ receptors. Clinical experience of the 5-HT$_2$/$_{1C}$ antagonist ritanserin has not, so far, revealed important anxiolytic effects but there are recent suggestions that the combined 2 and 1C antagonism may neutralize an anxiolytic effect mediated at one or other of the receptors (Ceulemans *et al.* 1985; Guimares *et al.* 1990). The drug may also have some non-5-HT-related functions which interfere with an anxiolytic effect.

THE MEDIAN RAPHE NUCLEUS AND CHRONIC AVERSIVE STIMULI

The median raphe nucelus innervates more cortical areas than the dorsal and its terminals parallel the distribution of noradrenergic terminals in the forebrain. Postsynaptic 5-HT$_{1A}$ receptors are highly concentrated in the hippocampus particularly in parts innervated by the median raphe nucleus. This system may have the function of coping with chronic aversive stimulation by progressively weakening the acute response to aversion when it is repeated. In this way normal behaviour emerges despite chronic adversity (Deakin 1988; Deakin and Graeff 1991).

This idea originates with the findings of Kennett *et al.* (1985) with their immobilization stress model of depression. This involves immobilizing animals by taping them to a wire grid for 2 h. 24 h later, when the animals are placed in an open field, they show considerably reduced locomotion compared with unstressed controls. Kennett *et al.* (1985) showed that with repeated immobilization stress the acute response disappeared—after seven daily immobilizations there was no difference between stressed and unstressed controls. Clearly some mechanism attenuated the acute effect of stress with repetition. They suggested 5-HT_{1A} receptors were involved because the animals became functionally supersensitive to 5-HT_{1A} agonists in behavioural assays as they became adapted to the stress. Furthermore, drugs which promote 5-HT_{1A} functioning when administered prior to, or immediately after, the first immobilization stress could accelerate the development of adaptation (Dickinson *et al.* 1985).

This finding may relate to an earlier formulation on the behavioural functions of the median raphe system (Deakin 1983). A number of experiments found that animals with damage to ascending 5-HT systems overrespond in a variety of experimental paradigms when further responding is no longer required. These include an extinction of appetitive behaviour, latent inhibition, habituation of exploration, and discrimination reversal (Deakin *et al.* 1979; Solomon *et al.* 1980). Conversely, drugs which enhance 5-HT neurotransmission have been reported to interfere with new learning in a variety of paradigms (Ogren and Johansson 1985; Winter and Petti 1987; Rowan *et al.* 1990). These findings suggest that the median raphe nucleus may function as a kind of disconnection or disengagement mechanism which normally serves to disconnect responses from the organisms behavioural repertoire when they are redundant or result in an aversive outcome. The concept of the median raphe system as a disconnection mechanism was strikingly corroborated by the finding that neurotoxin lesions of ascending 5-HT pathways actually improved acquisition of a visual discrimination task in rats (Altman *et al.* 1990). This finding is remarkable in that a lesion to the brain *improves* learning but the finding is predicted by the median raphe-disconnection theory with the additional assumption that the disconnection mechanism is tonically active.

Some form of disconnection or disengagement may mediate resilience in the face of chronic adversity. To think in human terms, imagine a woman with an alcoholic husband who is sometimes violent. This may engender a good deal of anxiety, but many women cope with this kind of chronic aversion without becoming depressed. Normal behaviour and the ability to enjoy social and other activities may continue despite the chronic aversion. Some mechanism disengages the chronic threat of the violent husband from influencing the rest of the woman's life. However, when this protective mechanism breaks down then depression results. In the following sections evidence is presented that the 5-HT_{1A} 'resilience' mechanism does breakdown in depression to cause the symptoms and the causes of the impairment are considered.

IMPAIRED 5-HT FUNCTION CAUSES DEPRESSION

Dr Cowen's chapter (Chapter 6) presents evidence that, indeed, in depression there is a breakdown of 5-HT_{1A} neurotransmission. The most compelling evidence comes from reproducible neuroendocrine abnormalities in depression (Cowen and Charig 1987; Deakin *et al.* 1990). Impaired 5-HT_{1A} neurotransmission may actually cause the state of depression because the nueroendocrine abnormalities are state dependent (Upadhyaya *et al.* 1991). Furthermore, all effective antidepressant drugs share the ability to enhance 5-HT_{1A} neurotransmission (de Montigny and Aghajanian 1984; Blier *et al.* 1987) and this appears to be necessary for their antidepressant efficacy (Delgado *et al.* 1990).

CAUSES OF IMPAIRED 5-HT_{1A} NEUROTRANSMISSION IN DEPRESSION

It is of considerable interest that environmental factors which are thought to predispose to depression have the ability in animal studies, and in some human studies, to reduce 5-HT_{1A} neurotransmission. Two important factors which increase the risk of depression are chronic psychosocial stress and social isolation (Brown and Harris 1978). In experiments at Manchester University, depressed patients had blunting of prolactin responses to intravenous infusions of the 5-HT precursor tryptophan (Deakin and Wang 1990). Blunting of the prolactin response occurred particularly in patients with high resting cortisol concentrations. The case notes of the patients were retrospectively analysed to determine whether psychosocial aetiological factors might be relevant to the neuroendocrine changes. Four factors were rated: antecedent life events, neurotic premorbid personality, chronic psychosocial stress, and duration of symptoms. Hypercortisolaemia was significantly associated with chronic psychosocial stress and chronic symptoms (Deakin and Wang 1990). These findings suggested the simple causal sequence:

psychosocial stress
↓
raised cortisol secretion
↓
impaired 5 HT function
↓
depression

A number of functional and biochemical investigations of 5-HT receptor neurotransmission show that glucocorticoids interfere with 5-HT_{1A} functioning and this would account for the blunting of prolactin responses to tryptophan in those with hyperecortisolaemia (Nausieda *et al.* 1982; Biegon *et al.* 1985; Dickinson *et al.* 1985; DeKloet *et al.* 1986; Bagdy *et al.* 1989). These findings suggest that chronic psychosocial stress may predispose to depression by causing hypercortisolaemia which interferes with 5-HT_{1A} neurotransmission—a system which is concerned with resilience and coping with chronic adversity.

Factors which increase social isolation predispose to depression and this, like psychosocial stress, may interfere with 5-HT_{1A} function. In animal experiments, housing animals in isolation is well known for its ability to cause marked behavioural and biochemical changes.

THE MEDIAN RAPHE DISCONNECTION MECHANISM AND SECONDARY DRIVE DISORDERS

The disengagement idea may be relevant to psychopathologies other than depression which have become associated with 5-HT. The various compulsive disorders are listed in Table 7.2. They are regarded as inappropriately

TABLE 7.2. *Secondary drive disorders.*

Alcoholism
Bulima
Gambling
Arson
OCD
Inapropriate learning, compulsions
Failure of/helped by MRN – 5-HT_{1A} disconnection

learned drives. The persistence of such compulsive behaviours could be due to some impairment of the median raphe mechanism which normally disengages inappropriate behaviours. Drugs which enhance 5-HT function may ameliorate these conditions by promoting disengagement or disconnection of inappropriate behaviour whether or not there is an underlying disturbance of 5-HT function.

CONCLUSION

This chapter argues that panic, anxiety, and depression can be seen as disorders of natural mechanisms of defence. They are disturbances in brain mechanisms which cope with aversive events. The wide range of psychopathologies associated with impaired 5-HT functioning and responsiveness to serotonergic drugs can be understood within this theoretical framework.

REFERENCES

Adler, c.A., Craske, G., and Barlow, D.H. (1987). Relaxation-induced panic (RIP): when resting isn't peaceful. *Integrative Psychiatry*, **2**, 94–112.

Altman, H.J., Normile, H.J., Galloway, M.P., Ramirez, A., and Azmitia, E.C. (1990). Enhanced spatial discrimination learning in rats following 5, 7-DHT-induced serotonergic deafferentation of the hippocampus. *Brain Research*, **518**, 61–6.

Audi, E.A., de Aguiar, J.C., and Graeff, F.G. (1988). Mediation by serotonin of the antiaversive effect of zimelidine and propranolol injected into the dorsal midbrain central grey. *Psychopharmacology*, **2**, 26–32.

Azmitia, E.C. and Segal, M. (1978). An autoradiographic analysis of the differential ascending projections of the dorsal and median raphe nuclei in the rat. *Journal of Comparative Neurology*, **179**, 641–68.

Bagdy, G., Calogero, A.E., Aulakh, C.S., Szemeredi, K., and Murphy, D.L. (1989). Long-term cortisol treatment impairs behavioural and neuroendocrine responses to 5HT1 agonists in the rat. *Neuroendocrinology*, **50**, 241–7.

Biegon, A., Rainbow, T.C., and McEwen, B.S. (1985). Corticosterone modulation of neurotransmitter receptors in rat hippocampus: a quantitative autoradiographic study. *Brain Research*, **332**, 309–14.

Blanchard, D.C. and Blanchard, R.J. (1972). Innate and conditioned reactions to threat in rats with amygdaloid lesions. *Journal of Comparative Physiological Psychology*, **81**, 281–90.

Blier, P., de Montigny, C., and Chaput, Y. (1987). Modifications of the serotonin system by antidepressant treatments: implications for the therapeutic response in major depression. *Journal of Clinical Psychopharmacology*, **7**, 24S–35S.

Blue, M.E., Yagaloff, K.A., Mamounas, L.A., Hartig, P.R., and Molliver, M.E. (1988). Correspondence between 5-HT2 receptors and serotonergic axons in rat neocortex. *Brain Research*, **453**, 315–28.

Bolles, R.C. and Fanselow, M.S. (1980). A perceptual-defensive-recuperative model of fear and pain. *Behavioural Brain Science*, **3**, 291–323.

Brown, G.W. and Harris, T.O. (ed.) (1978). *Depression: a study of psychiatric disorder in women*. Tavistock, London.

Ceulemans, D.L.S., M.L.J.A. Hoppenbrouwers, Gelders, Y.G. *et al*. (1985). The influence of ritanserin, a serotonin antagonist, in anxiety disorders: a double blind placebo-controlled study versus lorazepam. *Pharmacopysychiatry*, **18**, 303–5.

Costall, B., Naylor, R.J., and Tyers, M.B. (1990). The psychopharmacology of 5-HT$_3$ receptors. *Pharmacological Therapy*, **47**, 181–202.

Cowen, P.J. and Charig, E.M. (1987). Neuroendocrine responses to tryptophan in major depression. *Archives of General Psychiatry*, **44**, 958–66.

Critchley, M.A.E. and Handley, S.L. (1986). 5-HT2 receptor antagonists show anxiolytic-like activity in the x-maze. *British Journal of Pharmacology*, **89**, 646.

Davis, M. (1989). The role of the amygdaloid and its efferent projections in fear and anxiety. In *Psychopharmacology of anxiety* (ed. P. Tyrer), pp. 52–79. Oxford University Press.

Deakin, J.F.W. (1983). Roles of serotonergic systems in escape, avoidance and other behaviours. In *Theory in psychopharmacology*, Vol. 2 (ed. S.J. Cooper), pp. 149–93. Academic Press, New York.

Deakin J.F.W. (1988). 5HT2 receptors depression and anxiety. *Pharmacology Biochemical Behaviour*, **29**, 819–20.

Deakin, J.F.W. and Graeff, F.G. (1991). 5HT and mechanisms of defence. *Journal of Psychopharmacology*, **5**, 305–15.

Deakin, J.F.W. and Wang, M. (1990). Role of 5HT2 receptors in anxiety and depression. In *Serotonin: from cell biology to pharmacology and therapeutics* (ed. R. Paoletti, P.M. Vanhoutte, N. Brunello, and F.M. Maggi), pp. 505–9. Kluwer, The Netherlands.

Deakin J.F.W., File, S.E., Hyde, J.R.G., and Macleod, N.K. (1979). Ascending 5HT pathways and behavioural habituation. *Pharmacology Biochemical Behaviour*, **10**, 687–694.

Deakin, J.F.W., Pennell, I., Upadhyaya, A.K., and Lofthouse, R. (1990). A neuroendocrine study of 5HT function in depression: evidence for biological mechanisms of endogenous and psychosocial causation. *Psychopharmacology*, **101**, 85–92.

DeKloet, E.R., Sybesma, H., and Reul, M.H.M. (1986). Selective control by corticosterone of serotonin receptor capacity in raphe-hippocampal system. *Neuroendocrinology*, **42**, 513–21.

Delgado, P.L., Charney, D.S., Price, L.H., Aghajanian, G.K., Landis, H., and Heninger, G.R. (1990). Serotonin function and the mechanism of antidepressant action: reversal of antidepressant-induced remission by rapid depletion of plasma tryptophan. *Archives of General Psychiatry*, **47**, 411–18.

de Montigny, C. and Aghajanian, G.K. (1984). Tricyclic antidepressants: long-term treatment increases responsivity of rat forebrain neurones to serotonin. *Science*, **202**, 1303–6.

Den Boer, J.A. and Westenberg, H.G.M. (1990). Serotonin function in panic disorder: a double blind placebo controlled study with fluvoxamine and ritanserin. *Psychopharmacology*, **102**, 85–94.

Dickinson, S.L., Kennett, G.A., and Curzon, G. (1985). Reduced 5-hydroxy-tryptamine dependent behaviour in rats following chronic corticosterone treatment. *Brain Research*, **345**, 10–18.

Fernandez, A. and Hunsperger, R.W. (1962). Organization of the subcortical system governing defence and flight reactions in the cat. *Journal of Physiology*, **160**, 200–13.

Fibiger, H.C. and Phillips, A.G. (1988). Mesocorticolimbic dopamine systems and reward. *Annals of New York Academy of Science*, **537**, 206–15.

Goldberg, D.P., Bridges, K., Duncan-Jones, P. *et al.* (1987). Dimensions of neuroses seen in primary care settings. *Psychological Medicine*, **17**, 461–70.

Graeff, F.G. (1990). Brain defence systems and anxiety. In *Handbook of anxiety*, Vol. 3, (ed. M. Roth, G.D. Burrows, and R. Noyes), pp. 307–54. Elsevier, Amsterdam.

Guimares, F.S., Wang, M., and Deakin, J.F.W. (1990). Ritanserin reduces the expression of a conditioned fear response in neurotic patients. *Journal of Psychopharmacology*, **4**, 259.

Hodges, H., Green, S. and Glenn, B. (1987). Evidence that the amygdala is involved in benzodiazepine and serotonergic effects on punished responding but not in discrimination. *Psychopharmacology*, **92**, 491–504.

Jenck, E., Broekkamp, C.L.E. and Van Delft, A.M.L. (1989). Opposite control mediated by central 5HT1A and non-5HT1A (5HT or 5HT1C) receptors on periaqueductal gray aversion. *European Journal of Pharmacology*, **161**, 219–21.

Johnstone, E.C., Cunningham Owens, D.G., Frith, C.D., McPherson, K., Dowie, C., Riley, G. *et al.* (1980). Neurotic illness and its response to anxiolytic and antidepressant treatment. *Psychological Medicine*, **10**, 321–8.

Kalen, P., Rosegren, E., Lindvall, O., and Björklund, A. (1989). Hippocampal noradrenaline and serotonin release over 24 hours as measured by the dialysis technique in freely moving rats: correlation to behavioural activity state, effect of handling and tail-pinch. *European Journal of Neuroscience*, **1**, 181–8.

Kennett, G.A., Dickinson, S., and Curzon, G. (1985). Enhancement of some 5HT-dependent behavioural responses following repeated immobilization in rats. *Brain Research*, **330**, 253–63.

Kiser, R.S., Brown, C.A., Sanghera, M.K., and German, D.C. (1980). Dorsal raphe stimulation reduces centrally elicited fear like behaviour. *Brain Research*, **191**, 265–72.

Klein, D.F. and Klein, H.M. (1989). The definition and psychopharmacology of spontaneous panic and phobia. In *Psychopharmacology of anxiety* (ed. P. Tyrer), pp. 135–62. Oxford University Press.

Linnoila, M. and Virkkunen, M. (1991). Monoamines, glucose metabolism and impulse control. In *5-hydroxytryptamine in psychiatry: a spectrum of ideas* (ed. M. Sandler, A. Coppen, and S. Harnett), pp. 258–78. Oxford University Press.

Nashold Jr, B.S., Wilson, N.P., and Slaughter, G.S. (1969). Sensations evoked by stimulation in the midbrain of man. *Journal of Neurosurgery*, **30**, 14–24.

Nausieda, P.A., Carvey, P.M., and Weiner, W.J. (1982). Modification of central serotonergic and dopaminergic behaviours in the course of chronic corticosteroid administration. *European Journal of Pharmacology*, **78**, 335–43.

Ogren, S.-O. and Johansson, C. (1985). Separation of the associative and non-associative effects of brain serotonin released by *p*-chloroamphetamine: dissociable serotonergic involvement in avoidance learning, pain and motor function. *Psychopharmacology*, **86**, 12–26.

Pei, Q., Zetterström, T., and Fillenz, M. (1990). Tail pinch-induced changes in the turnover and release of dopamine and 5-hydroxytryptamine in different brain regions of the rat. *Neuroscience*, **35**, 133–8.

Rowan, M.J., Cullen, W.K., and Moulton, B. (1990). Buspirone impairment of performance of passive avoidance and spatial learning tasks in the rat. *Psychopharmacology*, **100**, 393–8.

Schutz, M.T.B., de Aguiar, J.C., and Graeff, F.G. (1985). Anti-aversive role of serotonin on the dorsal periaqueductal gray matter. *Psychopharmacology*, **85**, 340–5.

Shibata, K., Kataoka, Y., Gomita, Y., and Ueki, S. (1982). An important role of the central amygdaloid nucleus and mamilliary body in the mediation of conflict behaviour in rats. *Brain Research*, **372**, 159–62.

Solomon, P.R., Nichols, G.L., and Kaplan, L.J. (1980). Differential effects of lesions in medial and dorsal raphe of the rat: latent inhibition and septo-hippocampal serotonin levels. *Journal of Comparative Physiological Psychology*, **94**, 145–54.

Upadhyaya, A.K., Pennell, I., Cowen, P.J., and Deakin, J.F.W. (1991). Blunted growth hormone and prolactin responses to L-tryptophan in depression; a state-dependent abnormality. *Journal of Affective Disorders*, **21**, 213–18.

Walletschek, H. and Raab, A. (1982). Spontaneous activity of dorsal raphe neurons during defensive and offensive encounters in the tree-shrew. *Physiological Behaviour*, **28**, 697–705.

Winter, J.C. and Petti, D.T. (1987). The effects of 8-hydroxy-2-(di-*n*-propylamino) tetralin and other serotonergic agonists on performance in a radial maze: a possible role for 5HT1A receptors in memory. *Pharmacology, Biochemistry Behaviour*, **27**, 625–8.

8

Initial clinical psychopharmacological studies of α_2-adrenoceptor antagonists in volunteers and depressed patients

WILLIAM Z. POTTER, FRED GROSSMAN,
KARON DAWKINS, and HUSSEINI K. MANJI

In order to appreciate the interest in α_2-adrenergic-receptor antagonists in depression, it is useful to review relevant lines of evidence. Most of these reflect attempts to evaluate catecholamine, especially norepinephrine, function in depressed versus control subjects or to evaluate effects of antidepressant treatments on catecholamines.

A SYNTHESIS OF DATA IMPLICATING α_2-ADRENOCEPTORS IN DEPRESSION

Efforts to study norepinephrine output and function in affective disorders have evolved for almost two decades and yielded a cumulative database pointing to some overall degree of abnormality. Considerable evidence suggests that depressed patients excrete disproportionately greater amounts of norepinephrine and its major extraneuronal metabolite, normetanephrine, relative to total catecholamine synthesis (Manji *et al.*, in press). This is particularly true of melancholic, unipolar depressed subjects, but our recent data suggests that under adequately controlled drug-free (> 4 weeks) study, bipolar depressed subjects may exhibit a similar dysregulation of the noradrenergic system (Fig. 8.1). At least with regard to mania, the original catecholamine hypothesis has withstood the test of time, with increased noradrenergic function consistently observed in mania although this finding may ultimately reflect a secondary effect (Potter *et al.* 1987).

Findings of an increased fractional urinary output of urinary norepinephrine and normetanephrine and of an exaggerated rise in plasma norepinephrine upon orthostatic challenge (Rudorfer *et al.* 1985) in depressed patients are compatible with those of Esler *et al.* (1982) and Veith *et al.* (1988) of increased 'leakiness' of presynaptic NE terminals.

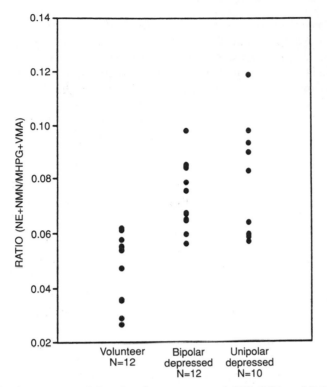

FIG. 8.1. A measure of fractional extraneuronal NE (NE + NM/MHPG + VMA) in 24 h urines from age- and sex-matched healthy control subjects and 3 week or more medication-free depressed unipolar and bipolar patients. NE = norepinephrine; NM = normetanephrine; MHPG = 3-methoxy-4-hydroxyphenylglycol; VMA = vanillylmandelic acid.

The mechanism(s) for this increased 'leakiness' remains unclear but may involve both central and peripheral sites.

One possible mechanism is a subsensitivity of noradrenergic nerve terminal α_2 autoreceptors. These receptors operate as 'thermostats' and depress norepinephrine release upon activation (Starke *et al.* 1989). Stimulation of these receptors does not directly influence the firing rate of the neuron but attenuates the release of neurotransmitters when an action potential depolarizes the varicosity or terminal. It has also been suggested that stimulation may even prevent the action potential from 'invading' the nerve terminal. Indeed, preclinical studies have demonstrated that blockade of presynaptic α_2 adrenoceptors on sympathetic neurones markedly increases the amount of neurotransmitter released per impulse (Starke *et al.* 1989). Subsensitive peripheral α_2 autoreceptors would be expected to result in a greater fractional excretion of norepinephrine and an exaggerated norepinephrine release upon any activation of the sympathetic nervous system (for example orthostatic stress, cold stress, 'early hospitalization stress', etc.). The suggestion of a subsensitivity of α_2 adrenoceptors receives additional support from findings in depression of blunted growth hormone responses to clonidine (reviewed by Siever 1987) and attenuated platelet α_2 function (as assessed by inhibition of AC activity) (Garcia-Sevilla *et al.* 1986). One might justifiably argue that subsensitive α_2 adrenoceptors may be the sequelae, not the cause, of increased norepinephrine release in depression. While this logically would apply to regulation of platelet (circulating) α_2 adrenoceptors, such a mechanism is difficult to reconcile teleologically for nerve terminal α_2 adrenoceptors in the central nervous system. If these autoreceptors do down regulate in the presence of increased norepinephrine, it would suggest the potential for an escalating feed-forward cycle, generally not seen in biological systems.

Another putative peripheral mechanism to explain the findings in depressed patients is that of an attenuation of the reuptake mechanism. The pattern of the relative excretion of norepinephrine and its metabolites in urine from depressed patients as compared to that from controls is strikingly similar to the further shift in the pattern observed during treatment with norepinephrine reuptake blockers (Linnoila *et al.* 1982). Although considerable evidence suggests that depression is associated with reduced platelet *serotonin* reuptake sites, this has not been adequately investigated for norepinephrine due to the absence of suitable peripheral tissues. There is some evidence that there may be physiologic regulation of the reuptake process (Lee *et al.* 1983), a possibility that could be studied further study in animal models or appropriate cultured cells. On the other hand, the failure to find alterations of infused ^{3}H-norepinephrine clearance in depression argues against a generalized abnormality of reuptake activity (Esler *et al.* 1982; Veith *et al.* 1988).

It is, of course, difficult to make inferences about 'brain norepinephrine' based on peripheral measures. Nevertheless, the increased sympathetic activity consistently observed in depression would be compatible with *increased* activity of the A6 locus coeruleus noradrenergic projections and/or *decreased* activity of the A1 'vasomotor' noradrenergic projections. The increased locus coeruleus activity would also be compatible with the observation of increased cerebrospinal fluid norepinephrine in depression and dysphoric mania (Post *et al.* 1984, 1989). Additionally, if one uses a 'pharmacologic bridge', the decreased firing of locus coeruleus neurones, the decreased levels of tyrosine hydroxylase and of its mRNA in locus coeruleus neurones (Nestler *et al.* 1990), and the reduced turnover in rat brain following chronic administration of antidepressants, would support such a contention (reviewed in Potter *et al.* 1988). Moreover, the observations by several independent laboratories that antidepressants, including the α_2-adrenoceptor antagonist, idazoxan, down regulate β adrenoceptors *in vitro* (in the absence of presynaptic inputs) (Hoenegger *et al.* 1986; Fishman *et al.* 1987; Manji *et al.* 1991, 1992) suggests that the aforementioned biochemical and electrophysiologic changes need not simply be 'compensatory adaptations' to chronic elevation of intrasynaptic norpinephrine.

The more traditional formulation of how altered α_2 adrenoceptors may play a role in depression was linked to findings that the therapeutic effect of chronic antidepressant treatments may result from a decrease in the number of central presynaptic α_2 adrenoceptors. The proposed result would be less inhibition of norepinephrine release and a corresponding relative increase in intrasynaptic norepinephrine (reviewed by Kafka and Paul 1986). As noted above, however, teleologically such a feed-forward cycle would be unusual in a biological system. Some investigators refined the hypothesis to include a relative failure of receptors to modulate in response to biochemical changes, rather than a static increase or decrease in receptor number (Siever and Davis 1985). The field has not yet reached the point of integrating available data into a formulation that addresses more specific issues such as whether alterations occur primarily at pre- or postsynaptic receptors or whether only specific central noradrenergic pathways are involved. With this background, we will briefly review the overall data from clinical radioligand binding of platelet adrenoceptors based on the possibility that platelets might provide a model for neurones. A more extended presentation and critique of this data is being published elsewhere (Grossman *et al.* 1993).

RADIOLIGAND BINDING STUDIES IN DEPRESSION

Data generated from radioligand binding studies in human platelets most commonly involve ^{3}H-yohimbine as a selective α_2-adrenoceptor antagonist and [^{3}H]-clonidine as a selective α_2-adrenoceptor agonist. The overall data using yohimbine generally does not yield significant differences between depressed patients and control groups for B_{max} or K_d. The range of mean values for B_{max} and K_d in depressed patients and control groups varies considerably (Table 8.1). The mean values across reports from different research settings for B_{max} range from 88 $\pm$ 45.1 fmoles/mg protein (Wolfe *et al.* 1986) to 347 $\pm$ 115.3 fmoles/mg protein (Horton *et al.* 1986) in depressed patients, and 124 $\pm$ 78.1 fmoles/mg protein (Wolfe *et al.* 1986) to 365 $\pm$ 16 fmoles/mg protein (Katona *et al.* 1989) in control groups, i.e. 3–4 fold. The mean values for K_d range from 0.95 $\pm$ 0.01 nM (Daiguji *et al.* 1981) to 5.2 $\pm$ 3.4 nM (Smith *et al.* 1983) in depressed patients and 0.92 $\pm$ 0.21 nM (Daiguji *et al.* 1981) to 4.0 $\pm$ 2.0 nM (Smith *et al.* 1983) in control

TABLE 8.1. *Yohimbine binding in depressed and control populations.*

Reference	Concentration of ligand	No. controls No. patients	Age controls Age patients	K_d controls (nM) K_d patients (nM)	B_{max} controls B_{max} patients (All but one study in fmol/mg protein)	Drug-free period	Diagnosis
Daiguji et al. 1981	[0.25–3.0 nM]	9 (UP) / 8 9 (BP) / 3 (BP)	26 ± 3 / 34 ± 14	0.92 ± .21 (UP) / 0.97 ± .23 0.92 ± .21 (BP) / 0.95 ± .01	240 ± 57 / 201 ± 76 240 ± 57 / 214 ± 24.3	7 days	Major depression Unipolar (UP)—8[a] Bipolar (BP)—3
Smith et al. 1983		16 / 18		4.0 ± 2 / 5.2 ± 3.4	165 ± 48 / 183 ± 38.2	14 days	Major depression
Stahl et al. 1983		29 / 16		2.1 ± 0.5 / 2.6 ± 0.4	180 ± 63[b] / 148 ± 52	28 days	UP—15 BP—1
Lenox et al. 1983	[1.0–10.0 nM]	18 / 20		1.91 ± 0.22[c] / NR[d]	289 ± 89[c] / NR		Primary dep.—85% Bipolar dep.—26%
Campbell et al. 1984	[0.5–20.0 nM]	14 / 12			228.9 ± 22 / 202.9 ± 12	21 days[e]	Major depression

Study						
Wolfe et al. 1986 [0.25–8.0 nM]	$\frac{18}{31}$	$\frac{37 \pm 11.7}{40 \pm 11.7}$	$\frac{1.47 \pm 0.63}{1.05 \pm 0.47}$	$\frac{124 \pm 78.1}{88 \pm 45.1}$	14 days[f]	Major depression[g]
Braddock et al. 1986	$\frac{39}{48}$	$\frac{32 \pm 10}{50}$	$\frac{3.5 \pm 0.8}{3.6 \pm 1.2}$	$\frac{43 \pm 13 \text{ fmol}/10^8 \text{ PLTS}}{45 \pm 13 \text{ fmol}/10^8 \text{ PLTS}}$	17 subj., 7 days[g] 34 subj., 14 days	Depressive illness
Horton et al. 1986	$\frac{41}{46}$		$\frac{2.26 \pm 1.22}{1.94 \pm 0.75}$	$\frac{362 \pm 96}{347 \pm 115.3}$	1–42 days[h]	Major depression[i]
Katona et al. 1989 [0.3–12.0 nM]	$\frac{44}{46}$	$\frac{40.6 \pm 11.9}{45.9 \pm 14.2}$	$\frac{2.33 \pm 1.13}{1.89 \pm 0.75}$	$\frac{365 \pm 106}{345 \pm 115}$	7 days	Major depression
Wolfe et al. 1989	$\frac{0}{26}$	$\frac{NR}{44.5 \pm 11}$	$\frac{NR}{1.1 \pm 0.5}$	$\frac{NR}{93.7 \pm 50.8}$	14 days[j]	Major depression[k]

SEM was converted to SD when data was available.

[a] 7 of 8 patients had delusions.
[b] Originally reported as sites/platelet—converted to fmol/mg protein from reported specific activities of the radioligands and on known values for the amount of proteins per platelet.
[c] Not reported.
[d] K_d and B_{max} for patients not given but reported as not significantly different from depressed group.
[e] Benzodiazepines given.
[f] 1 patient on furasamide, 1 patient on hydrochlorthiazide.
[g] Excluded patients who were suicidal.
[h] Benzodiazepines given.
[i] All but 4 patients diagnosed as having UP.
[j] 1 patient received 2–5 mg/day diazepam.
[k] Excluded patients who were suicidal.

groups, i.e. 4–5 fold. Some groups attributed the absence of positive findings, with yohimbine as an α_2-adrenoceptor antagonist, to its binding characteristics (Garcia-Sevilla *et al*. 1986; Kafka and Paul 1986), emphasizing an antagonist cannot reflect the properties of receptors in the high affinity state which may be more relevant to alterations in norepinephrine concentrations. Although use of an antagonist might explain negative results, it cannot explain the wide variability in mean values of B_{max} and K_d across studies.

Radioligand binding studies across groups are confounded by methodologic problems, such as varying patient composition, sex, age, clinical state, drug washout period, assay and platelet harvesting techniques, all of which have been extensively reviewed by others (Piletz *et al*. 1986). Another type of possible intrinsic methodological problem emerges from the observation that, under some circumstances, α_2-adrenoceptor 'desensitization' in human platelets following exposure to an agonist reflects nothing more than retained agonist which blocks binding sites, thus serving as a competitive inhibitor of additional agonist and perhaps antagonist as well (Karliner *et al*. 1981).

The investigators who raised this possibility designed a study to see if the decreased yohimbine binding demonstrated after preincubation of platelets with (−) epinephrine was the result of competition by epinephrine retained on receptors showing a high affinity for the agonist. Were this the case, then lowering the affinity of the receptor for the agonist (−) epinephrine by the addition of excess GTP (and Na^+) should reverse the decrease. They found that, in the presence of excess GTP (and Na^+), preincubation with (−) epinephrine no longer reduced yohimbine binding (Karliner *et al*. 1981). They thus supported the hypothesis that the apparent down regulation observed with platelets preincubated with (−) epinephrine was due to retained hormone on those receptors in the high affinity state. Whether this mechanism explains a significant portion of the variation in yohimbine binding results across studies has not been tested. Whatever the source(s) of variation, it is difficult to draw any conclusions as to the state of α_2 adrenoceptors in depressive disorders based on the available radioligand studies using yohimbine binding to platelets.

The overall data using clonidine as a radioligand for α_2 adrenoceptors on platelet membranes of depressed patients and control groups yield substantially different and more consistent results. Of seven studies reporting values for depressed patients and control groups, six report an increase in the mean values of B_{max} in depressed patients (Table 8.2). Three studies report findings consistent with a two-site model of binding which is described as reflecting 'high' and 'low' affinity states of the receptor. Two of these three studies reported significantly increased numbers of receptors in the high affinity state. The mean values for K_d yield less consistent findings.

Two of six studies reported increased mean values for K_d in depressed patients. The remaining four showed no significant differences. Only three studies separated the K_d values for 'high' and 'low' affinity states of the α_2 adrenoceptors. One reported an increase in mean K_d values for the 'low' affinity state receptors in depressed patients. Another reported increases in K_d for both the 'high' and 'low' affinity states of the receptors in depressed patients, while the third found no differences in mean K_d values.

In Table 8.2 we have also included two studies which did not report values in controls. A striking feature of the studies using clonidine as a ligand is the relatively narrow range across most groups for mean values of B_{max} and K_d, whether in depressed patients or normal controls. Excluding those studies which attempted to distinguish receptors in the high versus low affinity state, the mean values for B_{max} range from 38 fmole/mg protein (Campbell *et al.* 1984) to 56.9 ± 21.9 fmole/mg protein (Georgotas *et al.* 1987) in depressed patients and 31.6 ± 9.7 fmole/mg protein (Pandey *et al.* 1989) to 62.5 ± 24.5 fmole/mg protein (Georgotas *et al.* 1987) in control groups. Interestingly, the high B_{max} value in controls from Georgotas *et al.* (1987) is almost twice that in five out of the other six studies reviewed, while the B_{max} in depressed patients reported by this group is much closer to that reported by others (Table 8.2). The mean values for K_d (with the exclusion of studies with two affinity state measures) range from 4.5 ± 3.4 nM (Pandey *et al.* 1989) to 5.6 ± 3.2 nM (Smith *et al.* 1983) for depressed patients and 4.56 ± 2.84 nM (Georgotas *et al.* 1987) to 5.5 ± 3.1 nM (Smith *et al.* 1983) in control groups.

If general methodologic criticisms concerning intergroup data as elaborated for the studies using yohimbine were that relevant, then one might have expected more variation in the studies using clonidine. With the exceptions noted, the data from most of the studies using clonidine are remarkably consistent with little variability in the range of mean values (Table 8.2). There are, however, other limitations to interpreting findings with clonidine that go beyond reproducibility of measures across groups. First is the recognition, based on differences in ligand binding properties as well as more recent cloning studies, that α_2 adrenoceptors do not represent a homogeneous class (Bylund 1985; Kobilka *et al.* 1987; Lanier *et al.* 1988). Second, clonidine binds to sites that do not belong to any class of α_2 adrenoceptors as discussed in what follows.

IMIDAZOLINE RECEPTORS

Several lines of evidence suggest that the classes of drugs used as putative α_2-adrenoceptor ligands, imidazolines, and phenylethylamines, may act on two different types of presynaptic α_2 adrenoceptors (Ruffolo *et al.* 1977,

TABLE 8.2. *Clonidine binding in depressed and control populations.*

Reference	Concentration of ligand	No. controls / No. patients	Age controls / Age patients	K_d controls (nM) / K_d patients (nM) High (H) and low (L) if reported	B_{max} controls (fmol/mg protein) / B_{max} patients	Drug-free period	Diagnosis
Garcia-Sevilla et al. 1981	[1×10^{-9} – 6.4×10^{-8}M]	$\frac{21}{17}$	$\frac{37.2 \pm 16.5}{46 \pm 16.5}$	(H) $\frac{5.0 \pm 4.6}{5.5 \pm 2.9}$ (L) $\frac{17.2 \pm 11.5}{19.0 \pm 14}$	(H) $\frac{34.2 \pm 11}{45 \pm 12.4}$ (L) $\frac{76 \pm 22.9}{102 \pm 57.7}$	14 days	Major depression
Smith et al. 1982		$\frac{26}{29}$		$\frac{5.5 \pm 3.1}{5.6 \pm 3.2}$	$\frac{32 \pm 13.1}{46 \pm 8.9}$	14 days	Major depression
Garcia-Sevilla et al. 1983		$\frac{19}{5}$			$\frac{36 \pm 3}{50 \pm 4}$		Major depression
Campbell et al. 1984	[2.0–40.0 nM]	$\frac{14}{12}$			$\frac{NR^a}{38}$	21 days[b]	Major depression
Doyle et al. 1985	[2.0–64.0 nM]	$\frac{10}{13}$	$\frac{75.2 \pm 5.2}{74.8 \pm 5.9}$	(H) $\frac{6.6 \pm 2.4}{5.8 \pm 2.1}$ (L) reported as no significant diff.	(H) $\frac{56.3 \pm 18.9}{110.4 \pm 34.4}$ (L) reported as no significant diff.	14 days	Unipolar depression

Study	Concentration [ratio]				Duration	Diagnosis
Sacchetti et al. 1985	[1–8 × 10⁻⁹ M] $\frac{0}{31}$	$\frac{\text{NR}}{43.6 \pm 11.2}$	$\frac{\text{NR}}{4.4 \pm 3.4}$	$\frac{\text{NR}}{41.3 \pm 16.8}$	21 days	Major affective disorder
Garcia-Sevilla et al. 1986	[0.5 × 10⁻⁹ – 6.4 × 10⁻⁸ M] $\frac{20}{13}$	$\frac{35 \pm 17.9}{46 \pm 17.4}$	(H) $\frac{7.0 \pm 3.6}{13.9 \pm 8.3}$ (L) $\frac{27.5 \pm 26.4}{42.7 \pm 66}$	(H) $\frac{35 \pm 13.4}{53 \pm 10.8}$ (L) $\frac{63 \pm 17.9}{147 \pm 101}$	14–49 days	Major affective disorder[d]
Georgotas et al. 1987	[0.5–8.0 nM] $\frac{10}{14}$	$\frac{62.5 \pm 3.4}{62.5 \pm 5.5}$	$\frac{4.56 \pm 2.84}{5.53 \pm 2.92}$	$\frac{62.45 \pm 24.52}{56.86 \pm 21.90}$	7 days	Major depression[e]
Pandey et al. 1989	[0.5–35.0 nM] $\frac{33}{24}$ UP	$\frac{33.9 \pm 9}{41.2 \pm 12.2}$	$\frac{5.0 \pm 2.5}{4.5 \pm 3.4}$	$\frac{31.6 \pm 9.7}{46.3 \pm 26.5}$	14–21 days	Major depression UP = unipolar BP = bipolar
	$\frac{33}{24}$ BP	$\frac{33.9 \pm 9}{36.6 \pm 12.5}$	$\frac{5.0 \pm 2.5}{4.5 \pm 3.4}$	$\frac{31.6 \pm 9.7}{38.4 \pm 16.5}$		

SEM was converted to SD when data was available.
[a] Not reported.
[b] Benzodiazepines given.
[c] Outpatients.
[d] Outpatients—10 major depression; 3 bipolar.
[e] 11 patients 'endogenous'; 4 patients 'nonendogenous'.

1980; Langer 1980; Hicks *et al*. 1985). Furthermore, recently the existence of a distinct imidazoline receptor has become generally accepted. The α_2-adrenoceptor ligands with an imidazoline or imidazole structure (for example idazoxan, clonidine, *p*-aminoclonidine) bind to two or more sites in rodent, bovine, and human brain, an α_2 adrenoceptor and probably two different nonadrenoceptor sites, insensitive to catecholamines (Boyajian and Leslie 1987; Ernsberger *et al*. 1987; Bricca *et al*. 1988, 1989). Binding and auto-radiographic studies in rat brain have shown that [^{3}H]-idazoxan labels a heterogeneous population of binding sites, approximately four times the number of sites as does [^{3}H] rauwolscine, another α_2 antagonist. Thus, in contrast to the imidazoline compounds such as clonidine and idazoxan, the yohimbine diastereoisomers, yohimbine and rauwolscine, bind only weak-ly to the imidazoline receptor and therefore are more selective for the α_2-adrenoceptor (Parini *et al*. 1989).

Far from simply representing pharmacological binding sites, the imida-zoline sites appear to represent true functional receptors. Bousquet and colleagues have reported that the hypotensive effects of clonidine and related imidazolines may result from their interaction with imidazoline receptors in the nucleus reticularis lateralis rather than with α_2 adreno-ceptors (Bousquet *et al*. 1984; Tibirica *et al*. 1989). Thus, the hypotensive effects of a number of compounds hitherto considered selective for α_2-adrenoceptor sites appear to be due to their imidazoline moiety. In con-trast, since rauwolscine and yohimbine are able to antagonize clonidine-induced sedation, this may represent a 'true' α_2 adrenoceptor effect (Timmermans *et al*. 1981), which may be more systematically studied in humans.

Most relevant to the interpretation of results from clonidine binding is the demonstration that imidazoline sites exist on human platelets (Michel *et al*. 1990). Additionally, the ratio of α_2 adrenergic and imidazoline sites in human platelet membranes varies considerably between various donors (Michel *et al*. 1990). Although the aforementioned study was conducted using [^{3}H]-idazoxan, it appears very likely that clonidine (like idazoxan, an imidazoline compound) also binds to these nonadrenergic sites on human platelets. Intriguingly, the presence of the imidazoline sites on platelets may account for the observation of *two* high affinity sites (as well as a low affinity) on platelets, using the imidazoline agents, clonidine and *p*-aminoclonidine. Furthermore, a recent report suggests that a finding of elevated *p*-aminoclonidine binding in depression (the same as the finding with clonidine itself) is due solely to increases in the catecholamine-insensitive (imidazoline) sites (Piletz and Halaris 1991). Thus, platelet binding studies carried out to date cannot be confidently used to support the attractive hypothesis of altered α_2-adrenoceptor function in depres-sion. None the less, as described in what follows, studies in humans are being carried out based on the concept that manipulation of α_2 adrenoceptors may have therapeutic benefit.

EFFECTS OF PUTATIVE α_2-ADRENOCEPTOR ANTAGONISTS IN HUMAN VOLUNTEERS

Given uncertainty about the pathophysiologic basis of any abnormalities of noradrenergic systems in depression and α_2 receptors in particular, there is no direct argument for a therapeutic effect of α_2 antagonism in depression. Two indirect pharmacologic rationales, however, have led investigators to seriously consider such a possibility. First, as reviewed by Pinder and Sitsen (1987), it has been speculated that the antidepressant effects of mianserin are related to its antagonism of α_2 receptors, mainly because no other mechanism was identified. Second, the combination of α_2-adrenoceptor antagonists and norepinephrine uptake inhibitors leads to a greatly accelerated time course of β_1-adrenoceptor down regulation in rat brain (Crews et al. 1981; Keith et al. 1986). Since, β-adrenoceptor down regulation has been argued to be one of and perhaps the most common chronic biochemical effect of antidepressant treatments (Sulser et al. 1978), the potentiating effect of α_2 antagonism is highly suggestive. The first attempt to test the relevance of this phenomenon in humans was to add yohimbine to desipramine in patients who had not responded to the tricyclic alone (Charney et al. 1986). In this group of refractory patients, the addition of yohimbine was without benefit. Subsequently, more selective antagonists of α_2 receptors have been developed for human use, such as idazoxan, atipamezole, MK-912, and fluparoxan, which are characterized acutely in vivo as capable of blocking effects of various α_2 agonists (Elliot et al. 1984; Scheinin et al. 1988; Warren et al. 1989; Halliday et al. 1991). Of these, idazoxan and fluparoxan have been tried in some populations of depressed patients albeit with variable results (Osman et al. 1989; Corn, Chapter 14 of this book). The issue of clinical response to α_2 antagonists investigated to date will be discussed below. The proviso 'investigated to date' may be extremely important as regards the interpretation that the clinical effects of α_2 blockade have been adequately tested. It is possible that, with all tested compounds, effects other than simple α_2 blockade are playing a role. In what follows, the emphasis is on directly measurable chronic effects of drug and not on whether responses to administered α_2-adrenoceptor agonists have been altered. We are not aware of any attempt to date to relate the chronic drug effects to the degree of α_2-adrenoceptor blockade (assuming it were able to be established).

Chronic physiological, biochemical, and behavioural effects from administration of putative α_2-adrenoceptor antagonists are difficult to predict since most preclinical and clinical studies have been limited to measuring effects within the first 2–3 h following administration. For instance, there are a number of studies in rats which demonstrate 'positive' responses to idazoxan and/or yohimbine in behavioural paradigms that are distinguishable from those of amphetamine but can be broadly classified as stimulatory or enhancing some aspects of retention of learned behaviours (Table 8.3). These 'positive' effects are most marked when the α_2-adreno-ceptor antagonist is given 30 min prior to the behavioural test session and are lost if compound (for example idazoxan at 10 mg/kg intraperitoneally) is administered 120 min prior (Sanger 1988b). Similarly in terms of time course, clinical investigations have shown that acute yohimbine, idazoxan, and atipamezole produce variable increases from 100–600 per cent of, for instance, plasma norepinephrine that are maximal by 20–30 min following

TABLE 8.3. *Preclinical behavioural and 'cognitive studies' of α_2 antagonists*

Sanger 1988a	Idazoxan (0.6–10 mg/kg) and yohimbine (at 0.6–2.5 mg/kg but not at $\geq$ 5.0 mg/kg) increase bar pressing rates for food on a FI 60s schedule with a profile of behavioural stimulant effects distinct from those of amphetamine
Dickinson *et al.* 1989b	Acute idazoxan (1 mg/kg) given before but not after training improves passive avoidance responding, which suggests effects on perception/attention rather than memory/retrieval
Sanger 1989	Rats were trained to discriminate idazoxan (10 mg/kg IP) from saline. The discriminative stimulus was dose-related, generalized to yohimbine but not to the α_1 agonist cirazoline, α_1 antagonist prazosin, or α_2 agonist clonidine. The stimulus was not antagonized by prazosin or clonidine.
Sara and Devauges 1989	Acute idazoxan (2 mg/kg, i.p.), given to rats 4 weeks after maze training, alleviates 'forgetting'
Devauges and Sara 1990	Idazoxan (2 mg/kg i.p.) in rats had no effect in acquisition of a maze task or a visual discrimination task. Idazoxan did appear to facilitate the shift phase of the experiment and increased the amount of time spent on investigating novel stimuli
Dickinson and Gadie 1991	Idazoxan and RX 811059 induced reciprocal forepaw treading in acutely treated rats (po administration). This raises the possibility of direct or indirect 5-HT enhancement

intravenous infusion and return to baseline by 90 min (Elliott *et al.* 1984; Goldstein *et al.* 1991; Murburg *et al.* 1991; Scheinin *et al.* 1991). From the available data it is impossible to assess whether differences in the magnitude of increase of plasma norepinephrine (greatest after atipamezole) have most to do with differences in the specificity, potency and/or pharma-cokinetics of the compounds. Acute increases of blood pressure are con-sidered to be dose limiting so that full dose–response curves to establish the maximal norepinephrine response to each dose are not feasible in humans. Although of uncertain relevance to effects on peripheral cat-echolamines, idazoxan is reported to produce changes in the baseline EEG of awake rats that are not seen after the more selective α_2-adrenoceptor antagonists, efaroxan and ethoxyidazoxan (Dickinson and Gadie 1991). Such data may, however, be relevant to reports of differential effects of bedtime administration of 150 mg maprotiline, 0.10 mg clonidine, and 40 mg idazoxan *per os* (Nicholson and Pascoe 1991).

Against this background of transient acute effects, only a few psycho-pharmacologic studies have been focused on the consequences of chronic administration. In rats, the half-life of idazoxan is sufficiently brief that a method of continuous administration is necessary to study chronic effects. When given by osmotic minipump immediately after implant, idazoxan affected measured behaviour, an effect which disappeared with time. Con-versely, it had no early effect on cortical β_1 adrenoceptors and brain norepinephrine concentration in rats but after 10 days modestly decreased both without affecting peripheral norepinephrine measured in the atrium of the heart (Dickinson *et al.* 1990). These findings are consistent with a possible antidepressant effect of idazoxan since 10 but not 1, 3, or 7 days produced down regulation of cortical β adrenoceptors similar to that pro-duced by many other known antidepressants.

Published clinical studies on biochemical, physiologic, and behavioural effects following chronic administration of α_2 antagonists are only available for idazoxan in healthy volunteers. In an unusually comprehensive study, idazoxan was administered at a dose of 120 mg/day for three weeks to 12 normal male volunteers to assess effects on cortisol and ACTH, plasma 3-methoxy-4-hydroxy-phenylglycol (MHPG), melatonin, platelet and lymphocyte adrenoceptor binding, cardiovascular, sleep and behavioural parameters (Glue *et al.* 1991*a,b*, 1992; Wilson *et al.* 1991). Chronic idazoxan-reduced plasma MHPG while having no effect on nocturnal plasma melatonin, diastolic blood pressure, or heart rate (Glue *et al.* 1991*b*). Interestingly, there was a tendency for chronic idazoxan, opposite to its acute effects, to *reduce* baseline (pre-AM dose resting values) systolic blood pressure, suggesting 'an upregulation of central inhibitory α_2 adrenoceptors and subsequent reduction in noradrenergic activity' (Glue *et al.* 1991*b*). In these same subjects, chronic idazoxan tended to reduce total sleep and produced a significant decrease in REM sleep time accompanied by a marked increase in REM latency, a pattern of effects noted to be most similar to those observed after the MAOI class of antidepressants (Wilson *et al.* 1991).

Differences in acute versus chronic effects of idazoxan in these 12 volunteers were documented in two other areas. First, as reported after a single high dose (0.8 mg/kg) of yohimbine, acute idazoxan increased (prevented the clinical fall in) morning cortisol, an effect which was not observed following chronic administration (Glue *et al.* 1992). Interestingly, acute idazoxan is synergistic (not simply additive) with naloxone in stimulating plasma ACTH (Al-Damluji *et al.* 1990); studies of naloxone-stimulated ACTH after chronic idazoxan have not been reported. Second, chronic but not acute idazoxan increased the density of platelet α_2 adrenoceptors without producing consistent changes in lymphocyte β-adrenoceptor number (Glue *et al.* 1991*a*). Taken together, the results from these and other studies are consistent with acute central and peripheral noradrenergic effects of idazoxan followed by compensatory chronic changes as reflected in reduced total norepinephrine turnover (decreased MHPG), α_2-adrenoceptor changes and a return of both cortisol and cardiovascular regulation to normal (or below).

We have recently completed a protocol involving administration of 120 mg/day of idazoxan to healthy volunteers for 2-week period, looking at a somewhat different group of measures. Much of the biochemical data is to still be analysed. We did not observe any gross behavioural or cardiovascular changes, findings consistent with the observations of Glue *et al.* (1991*b*) described above. We have carried out a preliminary analysis of behavioural responses before and 4 h following an acute infusion of alprazolam at a dose (0.02 mg/kg) designed to detect baseline or benzodiazepine-induced alterations in cognitive function following chronic idazoxan. It should be noted that the dose of alprazolam employed produces transient sedation for 2–3 h. Both baseline and alprazolam-induced decrements in vigilance and Buschke recall were totally unaffected by chronic idazoxan (Figs 8.2 and 8.3) as was degree of sedation. These data indicate that whatever compensatory central nervous system changes which occur on idazoxan do not alter the pharmacodynamics of a potent benzodiazepine agonist, even with respect to sedation (Fig. 8.4). Thus, despite evidence of acute interactions between benzodiazepine agonists and catecholaminergic responses to both stress and α_2-adrenoceptor antagonists (Goldstein *et al.* 1991), under conditions of chronic treatment cascades of events following benzodiazepine and α_2-adrenergic receptor manipulations are either relatively independent or homeostatically 'balanced' to some pretreatment level. Taken together, studies of idazoxan in healthy volunteers show that, at doses producing subtle chronic alterations of noradrenergic systems, remarkably few physiologic or behavioural

Vigilance D' Means

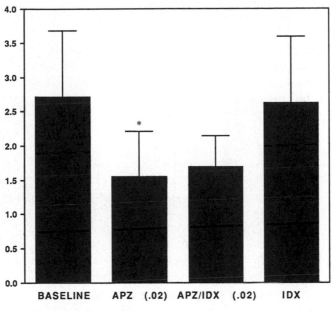

* P ≤ .05 (Baseline vs APZ)

FIG. 8.2. Lack of effect of chronic idazoxan (IDX) on alprazolam (APZ)-induced decrements in Vigilance D scores in six healthy volunteers. Subjects were tested for vigilance prior to (baseline) and 4 h after infusion of 0.02 mg/kg of APZ (second column), then again after 10 days of 40 mg t.i.d. IDX (last column) followed by a repeat APZ infusion (APZ/IDX, third column). Data from K. Dawkins, H. J. Weingartner, and W. Z. Pother (submitted).

effects are detectable. In fact, our subjects who received idazoxan under blind conditions could not detect above chance whether they were on drug, a finding consistent with the absence of behavioural effects on self-rated visual analogue scales (Fig. 8.5). Indeed, the only parameter tending to change was baseline anxiety which was in the direction of a decrease. This would be the expected direction of change over time in anxiety for volunteers participating in a research protocol involving repetition of procedures to which the subjects became more accustomed.

EFFECTS OF THE α_2-ADRENOCEPTOR ANTAGONIST, IDAZOXAN, IN PATIENTS

We previously reported our initial positive experience with idazoxan in three patients 'refractory' to standard antidepressants, two of whom were

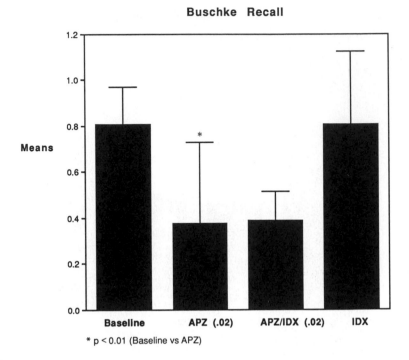

FIG. 8.3. Lack of effect of chronic idazoxan (IDX) in healthy volunteers on alprazolam (APZ)-induced decrements in Buschke Recall. Details of study as in Fig. 8.2 caption. Data from K. Dawkins, H. J. Weingartner and W. Z. Pother (submitted).

bipolar (Osman *et al.* 1989). Over the last 2 years we have carried out a 6–8 week trial of 80–120 mg/day of idazoxan alone or added to a stable dose of lithium in a mixed population of 13 bipolar I, II and unipolar patients, as well as one not otherwise classifiable depressed patient (Table 8.4). All patients received idazoxan under double-blind conditions during extended (minimum of four months) hospitalization. A positive clinical response was judged to be present if there was a three-point or greater reduction in the daily Bunney–Hamburg (B–H) (Bunney and Hamburg 1963) rating comparing the average of daily ratings for the week prior to beginning treatment to the average values following 4–6 weeks of treatment. For reference purposes, a total depression score of 6 on the B–H corresponds to a HDRS score of 25. A value of 2 or less on the B–H corresponds to remission (HDRS < 8). To date, as shown in Table 8.4, six patients (if one includes an apparent prophylactic response in subject No. 6) have responded. What was striking to us is that only 1/7 unipolar or nonbipolar patients responded versus 5/6 bipolars. We have therefore just initiated a

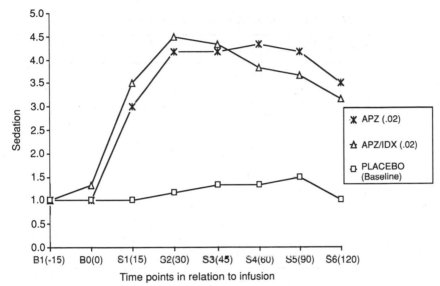

FIG. 8.4. Lack of effect of chronic idazoxan (IDX) in six healthy volunteers on alprazolam (APZ)-induced sedation. Absence of sedation after placebo prior to IDX treatment also shown. APZ infusion of 0.02 mg/kg over a 1-min period began at point BO. Overall protocol as described in Fig. 8.2.

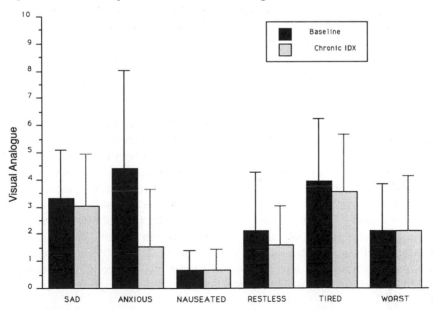

FIG. 8.5. Effects of chronic idazoxan (IDX) 40 mg t.i.d. × 10 days in six healthy volunteers on self-ratings using a 10 point visual analogue scale. Means ± SD are shown. No significant differences were observed. Unpublished data from Dawkins et al.

TABLE 8.4. *Inpatient idazoxan trial: 4–6 week responses*

Pt No.	Sex	DX	Response
1	F	BPI	No
2	M	UP	No
3	F	UP	Yes (with panic attack)
4	M	BPI	Yes
5	F	BPII	Yes
6	M	BPI	Prophylactic
7	F	UP	No
8	F	BPII	Yes
9	F	BPII	Yes
10	F	UP	No
11	F	UP	No
12	F	UP	No
13	M	NOS	No

prospective trial of idazoxan in bipolar patients among whom we will maintain those with a history of recurrent mania on lithium.

Our initial analysis of biochemical effects of idazoxan in the first six to ten of the patients listed in Table 8.4 provides evidence for compensatory changes in catecholaminergic measures after 4–6 weeks. Consistent with our initial observation in three patients, we observe increases of lying (mean ± SD pre-idazoxan 0.79 ± 0.48 pmol/ml; postidazoxan 1.67 ± 0.71 pmol/ml) and standing plasma norepinephrine (pre 2.16 ± 1.62 pmol/ml; post 3.22 ± 1.61 pmol/ml). And, consistent with Glue *et al.*'s (1991*b*) report of decreased free plasma MHPG in volunteers after idazoxan, total (free plus conjugated) MHPG tended ($p < 0.07$) to fall in urine (Table 8.5). What was most striking in the urinary measurements was that the ratio of norepinephrine + normetanephrine to MHPG + VMA increased in all but one subject investigated (Table 8.5 and Fig. 8.6). This ratio can be considered as a rough index of 'extraneuronal' norepinephrine since essential steps in the formation of MHPG and vanillylmandelic acid, which depend on deamination, occur primarily intraneuronally (Kopin 1985). The most obvious interpretation of the biochemical data in patients is that chronic blockade of α_2 adrenoceptors produced a modest decrease in centrally mediated sympathetic nervous system outflow while producing a relative enhancement of release of sympathetic nerve terminals (Manji *et al.* in press).

The point should be emphasized that we have included patients in the analysis who were on lithium; however, as shown in Fig. 8.6, at least with regard to the 'extraneuronal' norepinephrine ratio, all but one subject

TABLE 8.5. *Idazoxan: effects on urinary catecholamines (average of 2 days) following 4–6 weeks administration to depressed patients*

	Pre versus post		
	μM/24 h $\pm$ SD		*p*-value
Norepinephrine (NE)	0.67 $\pm$ 0.22	0.68 $\pm$ 0.18	> 0.10
Metanephrine ($N = 5$)	0.45 $\pm$ 0.18	0.51 $\pm$ 0.18	0.08
Normetanephrine (NM)	1.42 $\pm$ 0.60	1.61 $\pm$ 0.62	> 0.10
3-methoxy-4-hydroxyphenylglycol (MHPG)	9.78 $\pm$ 4.77	7.90 $\pm$ 3.1	0.07
Vanillylmandelic acid (VMA)	14.8 $\pm$ 5.5	12.4 $\pm$ 3.6	0.06
Homovanillic acid (HVA)	21.0 $\pm$ 4.2	23.0 $\pm$ 6.3	> 0.10
NE + NM/MHPG + VMA	0.086 $\pm$ 0.014	0.115 $\pm$ 0.018	< 0.001
HVA/MHPG + VMA	0.95 $\pm$ 0.36	1.14 $\pm$ 0.28	< 0.015

* $N = 10$ except as noted.

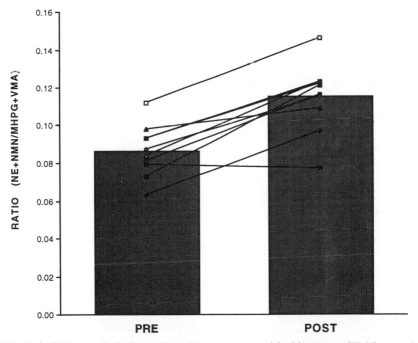

FIG. 8.6. Effects of 3–6 weeks of treatment with idazoxan (IDX) on the 'extraneuronal' ratio of NE in 24 h aliquots of urine obtained from 10 patients before and during treatment. Abbreviations are as defined in Fig. 8.1.

showed an increase. We would argue that the presence of lithium is not likely to alter the pattern of biochemical effects of idazoxan on absolute changes in norepinephrine turnover, metabolism and release despite our finding that lithium by itself (in normal volunteers) produces some similar changes (i.e. increases 'extraneuronal' norepinephrine). As we have discussed elsewhere with regard to the many potential sites of action of lithium distal to any specific receptor, there are several possible mechanisms whereby lithium might influence norepinephrine without directly altering α_2-adrenoceptor function (Risby *et al.* 1991).

SUMMARY

To date, clinical studies of overall noradrenergic function, attempts to measure α_2-adrenoceptor function, and initial therapeutic trials in depressed patients with available agents do not answer the question of whether there is some abnormality which can be beneficially modified by administration of α_2-adrenoceptor antagonists. None of the measures obtained can, with certainty, be attributed to alteration or manipulation of some specific class of α_2 adrenoceptor. Much of the data, however, is consistent with such a possibility. We are not discouraged at this point by the apparent limited efficacy of α_2 antagonists tried to date, which produce effects other than α_2 antagonism. Moreover, it is not clear that, with doses and schedules of administration so far tested, maximal tolerated α_2-adrenoceptor blockade in the central nervous system has been produced. Higher doses or more selective and potent drugs may reveal greater therapeutic efficacy. Furthermore, despite the initial negative data when yohimbine was combined with desipramine in refractory patients, it could still be that α_2-adrenoceptor antagonists will prove most effective when combined with other pharmacologic agents such as lithium which act at distinctly different molecular sites.

REFERENCES

Al-Damluji, S., Bouloux, P., White, A., and Besser, M. (1990). The role of alpha-2-adrenoceptors in the control of ACTH secretion; interaction with the opioid system. *Neuroendocrinology*, **51**, 76–81.

Bousquet, P., Feldman, J., and Schwartz, J. (1984). Central cardiovascular effects of alpha adrenergic drugs: differences between catecholamines and imidazolines. *Journal of Pharmacology and Experimental Therapeutics*, **230**, 232–6.

Boyajian, C.L. and Leslie, F.M. (1987). Pharmacological evidence for alpha2 adrenoceptor heterogeneity: differential binding properties of [^{3}H]rauwolscine and [^{3}H]idazoxan in rat brain. *Journal of Pharmacology and Experimental Therapeutics*, **241**, 1092–8.

Braddock, L., Cowen, P.J., Elliot, J.M., Fraser, S., and Stump, K. (1986). Binding of yohimbine and imipramine to platelets in depressive illness. *Psychological Medicine*, **16**, 765–73.

Bricca, G., Dontenwill, M., Molines, A., Feldman, J., Belcourt, A., and Bousquet, P. (1988). Evidence for the existence of a homogeneous population of imidazoline receptors in the human brainstem. *European Journal of Pharmacology*, **150**, 401–2.

Bricca, G., Dontenwill, M., Molines, A., Feldman, J., Belcourt, A., and Bousquet, P. (1989). The imidazoline preferring receptor: binding studies in bovine, rat and human brainstem. *European Journal of Pharmacology*, **162**, 1–9.

Bunney, W.E., Jr and Hamburg, D.A. (1963). Methods for reliable longitudinal observation of behavior. *Archives of General Psychiatry*, **9**, 114–28.

Bylund, D.B. (1985). Heterogeneity of alpha-2 adrenergic receptors. *Pharmacology, Biochemistry and Behavior*, **22**, 835–43.

Campbell, I.C., McKernan, R.M., Checkley, S.A., Glass, I.B., Thompson, C., and Shur, E. (1984). Characterization of platelet alpha$_2$ adrenoceptors and measurement in control and depressed subjects. *Psychiatry Research*, **14**, 17–31.

Charney, D.S., Price, L.H., and Heninger, G.R. (1986). Desipramine–yohimbine combination treatment for refractory depression. *Archives of General Psychiatry*, **45**, 1155–61.

Crews, F.T., Paul, S.M., and Goodwin, F.K. (1981). Acceleration of beta-receptor desensitization in combined administration of antidepressants and phenoxybenzamine. *Nature*, **209**, 787–9.

Daiguji, M., Meltzer, H.Y., Tong, C., U'Prichard, D.C., Young, M., and Kravitz, H. (1981). α_2-adrenergic receptors in platelet membranes of depressed patients: no change in number of ^{3}H-yohimbine affinity. *Life Sciences*, **29**, 2059–64.

Dawkins, K., Weingartner, H.J., Potter, W.Z. The effects of alprazolam and idazoxan on memory and sedation. Submitted.

Devauges, V. and Sara, S.J. (1990). Activation of the noradrenergic system facilitates an attentional shift in the rat. *Behavioural Brain Research*, **39**, 19–28.

Dickinson, S.L. and Gadie, B. (1991). Comparison of α_2-adrenoreceptor antagonists on background cortical EEG in the conscious rat. A role for the non-α_2-idazoxan-binding-site? *British Journal of Pharmacology*, **101**, 287.

Dickinson, S.L., Gadie, B., Havler, M.E., Hunter, C., and Tulloch, I.F. (1989*a*). Behavioural effects of idazoxan given continuously by osmotic minipump in the rat. *British Journal of Pharmacology*, **98**, 9 32P.

Dickinson, S.L., Gadie, B., and Tulloch, I.F. (1989*b*). Effect of idazoxan on passive avoidance behaviour in adult and in aged rats. *British Journal of Pharmacology*, **97**, 14P.

Dickinson, S.L., Gadie, B., Haynes, M.J., Lane, A.C., Walter, D.S., and Waltham, K. (1990). Neurochemical effects of idazoxan given continuously by osmotic minipump in the rat. *British Journal of Pharmacology*, **99**, 274P.

Doyle, M.C., George, A.J., Ravindran, A.V., and Philpott, R. (1985). Platelet α_2-adrenoreceptor binding in elderly depressed patients. *American Journal of Psychiatry*, **142**, 12.

Elliott, H.L., Jones, R., Vincent, J., Lawrie, C.B., and Reid, J.L. (1984). The alpha adrenoceptor antagonist properties of idazoxan in normal subjects. *Clinical Pharmacology Therapy*, **36**, 190–6.

Ernsberger, P., Meeley, M.P., Mann, J.J., and Reis, D.J. (1987). Clonidine binds to imidazole binding sites as well as a α_2-adrenoceptors in the ventrolateral medulla. *European Journal of Pharmacology*, **134**, 1–13.

Esler, M., Turbott, J., and Schwarz, R. (1982). The peripheral kinetics of norepinephrine in depressive illness. *Archives of General Psychiatry*, **39**, 295–300.

Fishman, P.H. and Finberg, J.P.M. (1987). Effect of the tricyclic antidepressant desipramine on β-adrenergic receptors in cultured rat glioma C6 cells. *Journal of Neurochemistry*, **49**, 282–7.

Garcia-Sevilla, J.A., Zis, A.P., Hollingsworth, P.J., Greden, J.F., and Smith, C. (1981). Platelet α_2-adrenergic receptors in major depressive disorder. Binding of tritiated clonidine before and after tricyclic antidepressant drug treatment. *Archives of General Psychiatry*, **38**, 1327–33.

Garcia-Sevilla, J.A., Garcia-Vallejo, P., and Guimon, J. (1983). Enhanced α_2-adrenoceptor-mediated platelet aggregation in patients with major depressive disorder. *European Journal of Pharmacology*, **94**, 359–60.

Garcia-Sevilla, J.A., Guimon, J., Garcia-Vallejo, P., and Fuster, M.J. (1986). Biochemical and functional evidence of supersensitive platelet α_2-adrenoreceptors in major affective disorder: effect of long-term carbonate treatment. *Archives of General Psychiatry*, **43**, 51–7.

Georgotas, A., Schweitzer, J., McCue, R.E., Armour, M., and Friedhoff, A.J. (1987). Clinical and treatment effects on ^{3}H-clonidine and ^{3}H-imipramine binding in elderly depressed patients. *Life Science*, **40**, 2137–43.

Glue, P., Payvandi, N., Kay, G., Elliott, J.M., and Nutt, D.J. (1991*a*). Effects of chronic α_2-adrenoceptor blockade of platelet and lymphocyte adrenoceptor binding in normal volunteers. *Life Sciences*, **49**, 21–5.

Glue, P., Wilson, S., Lawson, C., Campling, G.M., Franklin, M., Cowen, P.J. *et al.* (1991*b*). Acute and chronic idazoxan in' normal volunteers: biochemical, physiological and psychological effects. *Journal of Psychopharmacology*, **5**, 394–401.

Glue, P., Wilson, S., Campling, G., Knightly, M., Franklin, M., Cowen, P.J. *et al.* (1992). Alpha-2-adrenoceptor control of cortisol and ACTH in normal volunteers: effects of acute and chronic idazoxan. *Psychoneuroendocrinology*, **17**, 261–6.

Goldstein, D.S., Grossman, E., Listwak, S., and Folio, C.J. (1991). Sympathetic reactivity during a yohimbine challenge test in essential hypertension. *Hypertension*, **18** (Suppl. III), 40–8.

Grossman, F., Manji, H.K., and Potter, W.Z. (1993). Platelet alpha-2 adrenoreceptors in depression: a critical examination. *Journal of Psychopharmacology*, **7**, 4–18.

Halliday, C.A., Jones, B.J., Skingle, M., Walsh, D.M., Wise, H., and Tyers, M.B. (1991). The pharmacology of fluparoxan: a selective α_2-adrenoceptor antagonist. *British Journal of Pharmacology*, **102**, 887–95.

Hicks, P.E., Langer, S.Z., and Macrae, A.D. (1985). Differential blocking actions of idazoxan against the inhibitory effects of 6-fluoronoradrenaline and clonidine in the rat vas deferens. *British Journal of Pharmacology*, **86**, 141–50.

Hoenegger, U.E., Disler, B., and Wiesmann, U.N. (1986). Chronic exposure of human cells in culture to the tricyclic antidepressant desipramine reduces the number of beta-adrenoceptors. *Biochemical Pharmacology*, **35**, 1899–903.

Horton, R.W., Katona, C.L.E., Theodorou, A.E., Hale, A.S., Davies, S.L., Tunnicliffe, C. *et al.* (1986). Platelet radioligand binding and neuroendocrine challenge tests in depression, Antidepressant and receptor function. *Ciba Foundation Symposium*, **123**, 84–105.

Kafka, M.S. and Paul, S.M. (1986). Platelet α_2-adrenergic receptors in depression. *Archives of General Psychiatry*, **43**, 91–5.

Karliner, J.S., Motulsky, H.J., and Insel, P.A. (1981). Apparent 'down-regulation' of human platelet alpha$_2$-adrenergic receptors is due to retained agonist. *Molecular Pharmacology*, **21**, 36–43.

Katona, C.L.E., Theodorou, A.E., Davies, S.L., Hale, A.S., Kerry, S.M., Horton, R.W. *et al.* (1989). [^{3}H]Yohimbine binding to platelet α_2-adrenoceptors in depression. *Journal of Affective Disorders*, **17**, 219–28.

Keith, R.A., Howe, B.B., and Salama, A.I. (1986). Modulation of peripheral beta-1 and alpha-2 sensitivities by the administration of the tricyclic antidepressant, imipramine, alone and in combination with alpha-2 antagonist to rats. *Journal of Pharmacology and Experimental Therapeutics*, **236**, 356–63.

Kobilka, B.K., Matsui, H., Kobilka, T.S., Yang-Feng, T.L., Francke, U., Caron, M.G. *et al.* (1987). Cloning, sequencing, and expression of the gene coding for the human platelet alpha$_2$-adrenergic receptor. *Science*, **238**, 650–6.

Kopin, I.J. (1985). Catecholamine metabolism: basic aspects and clinical significance. *Pharmacological Reviews*, **37**, 333–64.

Langer, S.Z. (1980). Presynaptic regulation of the release of catecholamines. *Pharmacologic Reviews*, **32**, 337–62.

Lanier, S.M., Homcy, C.J., Patenaude, C., and Graham, R.M. (1988). Identification of structurally distinct alpha$_2$-adrenergic receptors. *Journal of Biological Chemistry*, **263**, 14 491–6.

Lee, C.M., Javitch, J.A., and Snyder, A. (1983). Recognition sites for norepinephrine uptake: regulation by neurotransmitter. *Science*, **220**, 626–9.

Lenox, R.H., Ellis, J.E., Van Riper, D.A., Ehrlich, Y.H., Peyser, J.M., Shipley, J.E. *et al.* (1983). Platelet α_2-adrenergic receptor activity in clinical studies of depression. In *Frontiers in neuropsychiatric research* (ed. E. Usdin), pp. 331–55. Macmillan Press, London.

Linnoila, M., Karoum, F., Calil, H.M., Kopin, I.J., and Potter, W.Z. (1982). Alteration of norepinephrine metabolism with desipramine and zimelidine in depressed patients. *Archives of General Psychiatry*, **39**, 1025–8.

Manji, H.K., Chen, G., Bitran, J.A., Gusovsky, F., and Potter, W.Z. (1991). Chronic exposure of C6 glioma cells to desipramine desensitizes β-adrenoceptors, but increases K_L/K_H ratio. *European Journal of Pharmacology: Molecular Pharmacology Section*, **206**, 159–62.

Manji, H.K., Bitran, J.A., Chen, G., Gusovsky, F., and Potter, W.Z. (1992). Idazoxan downregulates beta adrenergic receptors on C6 glioma cells *in vitro*. *European Journal Pharmacology: Molecular Pharmacology Section*, **227**, 275–82.

Manji, H.K., Rudorfer, M.V., and Potter, W.Z. (1994). Affective disorders and adrenergic function. In *Adrenergic dysfunction and psychobiology*. (ed. O.G.

Cameron). American Psychiatric Association Press, Washington, DC. (In press.)

Michel, M.C., Regan, J.W., Gerhardt, M.A., Neubig, R.R., Insel, P.A., and Motulsky, H.J. (1990). Nonadrenergic [^{3}H]idazoxan binding sites are physically distinct from alpha$_2$-adrenergic receptors. *Molecular Pharmacology*, **37**, 65–8.

Murburg, M.M., Villacres, E.C., Ko, G.N., and Veith, R.C. (1991). Effects of yohimbine on human sympathetic nervous system function. *Journal of Clinical Endocrinology and Metabolism*, **73**, 861–5.

Nestler, E.J., McMahon, A., Sabban, E.L., Tallman, J.F., and Duman, R.S. (1990). Chronic antidepressant administration decreases the expression of tyrosine hydroxylase in the rat locus coeruleus. *Proceedings of the National Academy of Science, USA*, **87**, 7522–66.

Nicholson, A.N. and Pascoe, P.A. (1991). Presynaptic alpha$_2$-adrenoceptor function and sleep in man: studies with clonidine and idazoxan. *Neuropharmacology*, **30**, 367–72.

Osman, O.T., Rudorfer, M.V., and Potter, W.Z. (1989). Idazoxan: a selective alpha-2 antagonist and effective sustained antidepressant in two bipolar depressed patients. *Archives of General Psychiatry*, **46**, 958–9.

Pandey, G.N., Janicak, P.G., Javaid, J.I., and Davis, J.M. (1989). Increased ^{3}H-clonidine binding in the platelets of patients with depressive and schizophrenic disorders. *Psychiatry Research*, **28**, 73–88.

Parini, A., Coupry, I., Graham, R.M., Uzielli, I., Atlas, D., and Lanier, S.M. (1989). Characterizations of an imidazoline/guanidinium receptive site distinct from the alpha$_2$-adrenergic receptor. *Journal of Biological Chemistry*, **264**, 11 874–8.

Piletz, J.E. and Hilaris, A. (1991). Noradrenergic imidazoline binding sites in platelets of depressed patients. *Biological Psychiatry*, **29**, 167A.

Piletz, J.E., Schubert, D.S.P., and Halaris, A. (1986). Evaluation of studies of platelet alpha$_2$ adrenoreceptors in depressive illness. *Life Sciences*, **39**, 1589–616.

Pinder, R.M., and Sitsen, J.M.A. (1987). α_2-adrenoceptor antagonists as antidepressants: the search for selectivity. In *Clinical pharmacology in psychiatry* (ed. S. Dahl, L. Gram, S. Paul, and W. Potter), pp. 107–12. Springer, Berlin.

Post, R.M., Jimerson, D.C., Ballenger, J.C., Lake, C.R., Uhde, W.W., and Goodwin, F.K. (1984). In *Neurology of mood disorders* (ed. R.M. Post and J.C. Ballenger), pp. 539–53. Williams and Wilkins, Baltimore.

Post, R.M., Rubinow, D.R., Uhde, T.W., Roy-Byrne, P.P., Linnoila, M., Rosoff, A. *et al.* (1989). Dysphoric mania. *Archives of General Psychiatry*, **46**, 353–8.

Potter, W.Z., Rudorfer, M.V., and Goodwin, F.K. (1987). Biological findings in bipolar disorders. In *American psychiatric association annual review*, Vol. 6 (ed. R.E. Hales and A.J. Frances), pp. 32–60. American Psychiatric Press, Washington, DC.

Potter, W.Z., Rudorfer, M.V., and Linnoila, M. (1988). New clinical studies support a role of norepinephrine antidepressant action. In *Prospectives in psychopharmacology: a collection of papers in honor of Earl Usdin* (ed. J. Barchas and D. Bunney), pp. 495–513. Alan R. Liss, New York.

Risby, E.D., Hsiao, J.K., Manji, H.K., Bitran, J., Moses, F., Zhou, D.F. *et al.*

(1991). The mechanisms of action of lithium. II. Effects on adenylate cyclase and beta receptor bindings in normals. *Archives of General Psychiatry*, **48**, 513–24.

Rudorfer, M.V., Ross, R.S., and Linnoila, M. (1985). Exaggerated orthostatic responsivity of plasma norepinephrine in depression. *Archives of General Psychiatry*, **42**, 1186–92.

Ruffolo, R.R., Turowski, B.S., and Patil, P.N. (1977). Lack of cross-desensitization between structurally dissimilar alpha-adrenoceptor agonists. *Journal of Pharmacy Pharmacology*, **29**, 378–80.

Ruffolo, R.R., Jr, Yaden, E.L., and Waddell, J.E. (1980). Receptor interactions of imidazolines. V. Clonidine differentiates postsynaptic alpha adrenergic receptor subtypes in tissues from the rat. *Journal of Pharmacology and Experimental Therapy*, **213**, 557–61.

Sacchetti, E., Conte, G., Pennati, A., Vila, A., Alciati, A., and Cazzullo, C.L. (1985). Platelet α_2 adrenoceptors in major depression: relationship with urinary 4-hydroxy-3-methoxy-phenylglycol and age at onset. Journal of Psychiatric Research, **19**, 579–86.

Sanger, D.J. (1988*a*). Behavioural effects of the α_2-adrenoceptor antagonist idazoxan and yohimbine in rats: comparisons with amphetamine. *Psychopharmacology*, **96**, 243–9.

Sanger, D.J. (1988*b*). The α_2-adrenoceptor antagonists idazoxan and yohimbine increase rates of DRL responding in rats. *Psychopharmacology*, **95**, 413–17.

Sara, S.J. and Devauges, V. (1989). Idazoxan, an α-2 antagonist, facilitates memory retrieval in the rat. *Behavioural and Neural Biology*, **51**, 401–11.

Scheinin, H., MacDonald, E., and Scheinin, M. (1988). Behavioural and neurochemical effects of atipamezole, a novel α_2-adrenoceptor antagonist. *European Journal of Pharmacology*, **151**, 35–42.

Scheinin, M., Karhuvaara, S., Ojala-Karlsson, P., Kallio, A., and Koulu, M. (1991). Plasma 3,4-dihydroxyphenylglycol (DHPG) and 3-methoxy-4-hydroxy-phenylglycol (MHPG) are insensitive indicators of α_2-adrenoceptor mediated regulation of norepinephrine release in healthy human volunteers. *Life Sciences*, **49**, 75–84.

Siever, L.J. (1987). Role of noradrenergic mechanisms in the etiology of the affective disorders. In *Psychopharmacology, the third generation of progress* (ed. H.Y. Meltzer), pp. 493–504. Raven Press, New York.

Siever, L.J. and Davis, K.L. (1985). Overview: toward a dysregulation hypothesis of depression. *American Journal of Psychiatry*, **142**, 1017–31.

Smith, C.B., Hollingsworth, P.J., Garcia-Sevilla, J.A., and Zis, A.P. (1983). Platelet alpha$_2$ adrenoreceptors are decreased in number after antidepressant therapy. *Progress in Neuropsychopharmacology and Biological Psychiatry*, **7**, 241–7.

Stahl, S.M., Lemoine, P.M., Ciaranello, R.D., and Berger, P.A. (1983). Platelet alpha$_2$-adrenergic receptor sensitivity in major depressive disorder. *Psychiatry Research*, **10**, 157–64.

Starke, K., Gothert, M., and Kilbinger, H. (1989). Modulation of neurotransmitter release by presynaptic autoreceptors. *Physiological Reviews*, **69**, 864–989.

Sulser, F., Vetulani, J., and Mobley, P.L. (1978). Mode of action of antidepressant drugs. *Biochemical Pharmacology*, **27**, 257–71.

Tibirica, E., Mermet, C., Feldman, J., Gonon, F., and Bousquet, P. (1989). Correlation between the inhibitory effect on catecholaminergic and entrolateral medullary neurons and the hypotension evoked by clonidine: a voltammetric approach. *Journal of Pharmacology and Experimental Therapeutics*, **250**, 642–7.

Timmermans, P.B., Schoop, A.M., Kwa, H.Y., and Van Zwieten, P.A. (1981). Characterization of alpha adrenoceptors participating in the central hypotensive and sedative effects of clonidine using yohimbine, rauwolscine and corynanthine. *European Journal of Pharmacology*, **70**, 7–15.

Veith, R.C., Barnes, R.F., Villacres, E., Marburg, M.M., Raskind, M.A., and Borson, S. (1988). Plasma catecholamines and norepinephrine kinetics in depression and panic disorder. In *Progress in catecholamine research, part c: clinical aspects* (ed. R. Belmaker, H. Sandler, and M. Dahlstrom), pp. 197–202. Alan R. Liss, New York.

Warren, J.B., Dollery, C.T., Fuller, R.W., Williams, V.C., and Gertz, B.J. (1989). Assessment of MK-912, an α_2-adrenoceptor antagonist, with use of intravenous clonidine. *Clinical Pharmacology and Therapeutics*, **46**, 103–9.

Wilson, S.J., Glue, P., and Nutt, D.J. (1991). The effects of the α_2-adrenoceptor antagonist idazoxan on sleep in normal volunteers. *Journal of Psychopharmacology*, **5**, 105–10.

Wolfe, N., Cohen, B.M., and Gelenberg, A.J. (1986). Alpha$_2$-adrenergic receptors in platelet membranes of depressed patients: increased affinity for ^{3}H-yohimbine. *Psychiatry Research*, **20**, 107–16.

Wolfe, N., Gelenberg, A.J., and Lydiard, R.B. (1989). Alpha$_2$-adrenergic receptor sensitivity in depressed patients: relation between ^{3}H-yohimbine binding to platelet membranes and clonidine-induced hypotension. *Biological Psychiatry*, **25**, 382–92.

9

Recurrent brief depression

STUART A. MONTGOMERY

In the absence of obvious aetiological or biochemical factors the diagnosis of depression has had to be based on the presence of specific psychopathology. In the classificatory systems that have been most recently developed, the diagnosis of major depression depends on the presence of dysphoric mood accompanied by a minimum number of defined depressive symptoms. The operational criteria for the diagnosis distinguish the syndrome of depression from normal, transient unhappiness that may be accompanied by a few depressive symptoms.

The distinction between depression requiring treatment and transient mood swings is also determined by the requirement of a minimum duration of the disorder included consistently in currently used diagnostic criteria. Some criteria, for example those of Feighner *et al.* (1972), require a longer period of four weeks but the DSM–III, III-R (APA 1980, 1987), the Research Diagnostic Criteria of Spitzer *et al.* (1978) (RCD), and the International Classification of Diseases (ICD-10) (WHO 1992) all require a minimum duration of 2 weeks.

The syndrome of major depression defined by these diagnostic criteria identifies a group with functional impairment who are likely to have a good response to conventional antidepressants. Research interest has tended to focus on this group of major depression since the introduction of the tricyclic antidepressants. What the diagnostic criteria overlook is the existence of a group who may have the same level of impairment as major depression but whose depressive episodes are of less than two weeks duration.

Until recently systematic investigation of depressive states lasting less than two weeks was almost entirely lacking. Failure to recognize their existence cannot explain the paucity of studies. Kraepelin included a category for milder, rapidly resolving forms of depression in his classification and there are numerous reports in the literature that suggest a relatively high incidence of brief depression. Paskind, for example, reported in 1929 that some 14 per cent of a large series of 663 inpatients suffered from a brief episode of depression lasting from a few hours to a few days. Brief depressions have also been commented on by earlier clinicians, for example Gregory (1915), Kraepelin, (1921), Bleuler (1924).

These brief episodes of depression may have received less attention because of an assumption that shorter equates with milder. This notion is implicit in the RDC, a diagnostic system which comes the nearest to recognizing brief depressions under its category of intermittent depression, but which defines this as a minor depression with few symptoms.

The importance of these brief episodes of depression was recognized more recently. Clinical studies reported the significance of depressive symptoms of short duration as predictors of suicide attempts (Montgomery et al. 1979, 1983), and epidemiological studies reported the prevalence of frequently occurring brief depressive episodes to be at least as high as major depression in the normal population (Angst and Dobler-Mikola 1985; Angst 1990).

The prevalence of a disorder will remain somewhat uncertain until it is clearly included within a generally accepted classificatory system so that systematic studies can be carried out. Recurrent brief depression (RBD) has, however, now been recognized in ICD-10 (WHO 1992) and studies have been carried out using this instrument as the basis of diagnosis which helps to characterize the disorder and redress the curious neglect of the condition. The high prevalence of brief depression is reported in epidemiological studies both in general practice (Lecrubier et al. 1992) and in psychiatric practice (Philipp and Benkert 1992) and these studies confirm the finding in the general population that RBD has a similar prevalence to major depression.

SEVERITY OF BRIEF DEPRESSION

The presenting psychopathology of the depressive episodes is the same as that for major depression. The same core symptoms of depression occur in both and the brief depressions have at least as many symptoms as major depression. The distinction between brief and major depression can only be made on the basis of duration with major depression lasting 2 weeks or more. A series of studies in psychiatric practice of patients complaining of brief depressions demonstrated that this is not a mild disorder (Montgomery et al. 1983, 1989, 1990; Montgomery and Montgomery 1992). Taking all the episodes together, whether mild, moderate, or severe, the mean depression severity score measured on the Montgomery and Asberg Depression Rating Scale (MADRS) (Montgomery and Asberg 1979) was 30.3 in the study of Montgomery et al. (1989). This level of severity is somewhat higher than that reported in some studies of the efficacy of antidepressant treatment in major depression from which mild depression has been excluded.

In this series of patients some 70 per cent of episodes were of moderate or greater severity which makes the separation of these brief depressions from the so-called minor depressions abundantly clear. The levels of severity observed in the first series of patients have now been replicated in subsequent series from prospective controlled studies (Montgomery and Montgomery 1992).

It may not be entirely fair to conclude that the brief depressions are more severe than major depression. Patients who are more severely ill are often excluded from controlled studies of major depression because it is recognized that reference tricyclic antidepressants may not be particularly effective in treating severe depression. Where a placebo control is included it is also possible that the more severe major depression may be excluded from placebo controlled studies. The effect of these exclusions may be to lower the mean severity scores of major depression group entering in prospective efficacy studies. Nevertheless the data from the studies of RBD are sufficient to support the notion that these brief depressions are at least of comparable severity to major depression.

DURATION OF EPISODES OF BRIEF DEPRESSION

The episodes of brief depression are not merely shorter than major depression, they are typically very short with a median duration recorded in the first London series of 3 days (Montgomery *et al.* 1989). Two-thirds of the episodes lasted between 2 and 4 days and 75 per cent of episodes lasted less than 4 days. This series included a group of patients with combined depression who occasionally developed episodes lasting more than 2 weeks. If these were excluded 81 per cent had an episode lasting 4 days or less. Very similar durations of the brief depressions were observed in subsequent series (Montgomery *et al.* 1990; Montgomery *et al.* 1992).

The 3-day period seen in the clinical studies is consistent with the results from epidemiological studies in the normal population. In Zurich, where a 10 year follow-up study has been carried out, the brief depressions are also reported to last 2–3 days (Angst *et al.* 1990). The results coming from the studies of the incidence and characteristics of brief depressions in general practice patients are also in accord (Lecrubier *et al.* 1992). In this sample the mean duration was reported to be around 3–4 days with a slightly higher number of patients reporting episodes that lasted between 1 and 2 weeks than were seen in the other studies.

The length of episodes reported in studies carried out in different countries and in different types of population samples are in substantial agreement.

It appears relatively easy to separate brief depression from major depression on the basis of duration taking the commonly used cut off of more than 2 weeks as a criterion to define major depression. For any individual patient with current depression it is relatively simple to ask whether a particular episode lasted less than 2 weeks.

Identifying recurrent brief depression in the history may sometimes be confounded by the distorting effects of memory. While there is evidence to suggest that more severe symptoms are remembered with greater clarity than milder episodes, memory of events over a long time span is likely to be faulty (Jenkins *et al.* 1979; Brown and Harris 1982). Brief episodes recurring with only short intervals between are sometimes coalesced and remembered as a single episode when recalled after a long period. Nor is it reasonable to expect a precise memory of the duration and severity of each episode after a few months in those who have frequent episodes. Such factors have undoubtedly contributed to the underestimation of the importance of brief depressive episodes.

INTERVALS AND RECURRENCE RATE

In the prospective clinical studies the brief depressive episodes recurred frequently over the period of observation. The episodes occurred erratically and were not regular or cyclical in recurrence. Irregularly regular would be a more fitting description. In the first series (Montgomery *et al.* 1989) the mean interval between the beginning of one episode and the start of the next was 18 days. The intervals lasted mostly between one and 5 weeks with only 14 per cent longer than 5 weeks. Some 25 per cent of patients have a period of more than 4 weeks between one episode and the next during the period of observation. Similar results are seen in the study in general practice in France where a substantial group had intervals between episodes of 2 months or more (Lecrubier 1992). This makes it plain that a strict monthly criterion of recurrence suggested in the Zurich criteria (Angst *et al.* 1988) is inappropriate.

The mean recurrence rate, estimated on the basis of the clinical studies, is approximately 20 episodes per year. This high frequency is supported by the Zurich epidemiological study which reported substantial occupational morbidity associated with 12 or more episodes per year. Larger databases are needed before the case can be made for a precise defining minimum number of episodes for the diagnosis. In the Zurich study the number of patients experiencing between 8 and 12 episodes a year was too small to analyse separately so that this data set cannot be used to define any minimum limit.

BRIEF DEPRESSION AND SUICIDAL BEHAVIOUR

Suicide and suicidal behaviour are both associated with RBD and are indicative of the seriousness of the condition. The clearcut association of suicidal behaviour and brief depressions lasting less than 2 weeks was identified in a series of placebo controlled studies investigating the efficacy of treatments in reducing suicidal behaviour. These 6-months prophylactic studies were conducted in a group of repeated suicide attempters from which major depression had been excluded (Montgomery *et al*. 1979, 1983; Montgomery and Montgomery 1982). Core depressive symptoms lasting less than 2 weeks seen in the first month of the studies predicted subsequent suicide attempts during the studies. The total score of the depression scale used, the MADRS, was a significant predictor of further suicidal behaviour as were six out of ten individual items on the scale (Montgomery *et al*. 1983). Suicide attempts in this group of recurrent attempters were apparently confined to the periods of brief depression, an impression that was confirmed in later close follow up studies in a similar cohort of patients.

It could be argued that the association between suicidal behaviour and brief depressive episodes was attributable to selection bias. However, confirmation of the finding is provided in the epidemiological studies. In the Zurich studies the suicide attempt rate was reported to be significantly higher in the brief depression group compared with the normal population group (Angst 1990). In a different population sample the suicide attempt rate was twice as high in the brief depression group compared with the major depression group although this difference was not significant. A more morbid group appears to include those individuals with combined depression who have a pattern of both major and brief depression. In this group there is a substantially increased suicide attempt rate over either RBD or major depression alone.

There are a number of factors which may contribute to the raised suicide risk in those who suffer recurrent brief depressive episodes. The rapid transition from normal to the depressed state is reported by many patients as difficult to tolerate and would obviously be likely to cause particular problems in those who develop severe depressions with marked suicidal thoughts. The unpredictability of the onset of the depressions also appears to increase the difficulty of developing coping strategies. There is also an increase in aggression and hostility during the brief depressive episode which may exacerbate the difficulties experienced. Although a similar increase in aggression is seen in major depression (Angst *et al*. 1990), there may be differences, for example in the levels of impulsivity, which increase the difficulty in coping with rapid changes in the severity of depression.

The greatly increased risk of suicide attempts in the combined depression group may merely identify a more morbid group. However, it seems possible that the suicidal behaviour may be mediated by different mechanisms in major depression and brief depression. In the Zurich epidemiological study 30 per cent of the group who developed both major depression and RBD had already attempted suicide by the age of 28 (Angst *et al.* 1990) making the presence of combined depression the best predictor of suicidal behaviour available to the clinician. The combination of lengthy periods of depression and the rapid onset of frequent three day depressions appears to be particularly dangerous and it is possible that different underlying mechanisms have an additive effect when they occur together.

COMBINED DEPRESSION

Some patients who have a clear pattern of recurring brief depressive episodes have occasional episodes of longer duration. In a follow up of patients with brief depressions in our psychiatric practice 8 per cent developed episodes of depression lasting 14 days or longer. Combined depression is the term chosen for those who suffer with major depression and recurrent episodes of brief depression (Montgomery *et al.* 1989).

An intervening episode of major depression does not appear to change the underlying pattern of RBD. When the major depression resolves the RBDs return. The brief depressions in this group are of the same approximate duration, with a median of 3 days, though there may be more episodes that last between 1 and 2 weeks and a slightly longer mean duration of the episodes of brief depression compared with those with the pure RBD.

The greater morbidity of the combined depression group is seen in the increased risk of suicide and also in the high rates of substance and alcohol abuse. Substance abuse was raised compared with the general population in both RBD and major depression but it was significantly higher in the combined depression group. Panic also occurs with a significantly higher frequency in the combined depression group (Angst *et al.* 1990).

STABILITY OF THE DIAGNOSIS

RBD is a stable condition with a large proportion of individuals with the condition continuing to suffer from brief episodes which recur over many years. In the 10-year follow-up in the Zurich cohort 56 per cent continued with episodes of brief depression compared with 60 per cent with major depression (Angst *et al.* 1990). There were in this study substantial numbers who developed combined depression. It does not appear from the epidemiological studies that RBD represents a prodromal phase of major depression. The 10-year follow up in a community sample shows that as many individuals switched from a diagnosis of pure RBD to combined depression as from pure major depression to combined depression. Furthermore the diagnosis of RBD had the same stability over the 10-year period as major depression (Angst *et al.* 1990). The stability of the anxiety disorders in this series was by contrast quite low with a greater proportion of individuals switching to a diagnosis of depression than remained as anxiety disorders (Vollrath and Angst 1989). This stability of the diagnosis of RBD has been demonstrated in a different psychiatric outpatient population in a one-year study (Philip and Benkert 1992).

IS RBD RELATED TO BIPOLAR DISORDER?

There are some parallels between RBD and rapid cycling. Both conditions have erratic recurrences of brief episodes of depression and both conditions appear not to respond to conventional antidepressants. Rapid cycling patients suffer episodes of mania or hypomania as well as episodes of depression. In contrast, the Zurich epidemiological study reported fewer episodes of mania or hypomania in those with RBD than are seen in the general population; on the other hand episodes of mania or hypomania were seen significantly more frequently in those with major depression than the general population. This strongly suggests that RBD is not related to bipolar illness. Our clinical data are in accord with the epidemiological data suggesting that RBD is not related to bipolar illness. There remains the possibility that RBD is the unipolar version of bipolar rapid cycling.

PSYCHOPHARMACOLOGY

The epidemiological studies suggest that a substantial proportion of those with RBD attend for treatment though quite what treatment they receive has yet to be identified. Many of those with brief depressions give a history of inadequate response to the treatments offered in primary care and of perplexity among treating psychiatrists in secondary or tertiary referral.

The nature of RBD with its intermittent bursts of quite severe but short illness makes it difficult to establish whether treatments are effective. Currently available antidepressants are slow to produce an effect and are therefore inappropriate to treat episodes of depression which last on average only 3 days. Only treatments with immediate efficacy could be expected to be useful in treating the individual episodes, in the same way, for example, that individual episodes of migraine can be treated.

Strategies for creating a new range of potential treatments aimed at rapid reversal of the depressed state are now needed and these treatments will require new strategies for testing, both in the laboratory and in the clinic. Many of the animal models of depression are behavioural, indirect, and may not be capable of detecting rapid onset of efficacy. New approaches are needed in the creation of animal models which can focus on rapid onset reversal of depression, for example tryptophan depletion or interferon-overload induced dysphoria.

In the absence of treatments with rapid onset of action the alternate strategy is to use treatments prophylactically. Long term treatment strategies have proved to be successful in treating major depression and a range of antidepressants have been shown to be effective in reducing the risk of subsequent depressive episodes (Montgomery and Montgomery 1992).

The designs used to assess antidepressants in the long-term treatment of major depression have been adapted for investigating potential prophylactic treatments for RBD. The conventional designs for investigating long-term treatment of major depression have concentrated on measuring the relapses and time to relapse of the single next episode in large groups of patients compared with controls. This is unlikely to be the most appropriate way of approaching an illness with rapid erratic recurrences. In some cases the recurrence may occur before the treatment has had a chance to exercise prophylactic action which would greatly reduce the power of the study to detect efficacy. The investigational designs have to be tailored to the nature of the illness and for RBD the most fruitful approach is likely to be a careful prospective analysis of the severity, duration, and recurrence rate, in a group of patients with established RBD.

CLINICAL INVESTIGATIONS OF RBD

Since the World Health Organization included RBD in the International Classification of Diseases (tenth revision) (WHO 1992) the impetus to investigate potential efficacy of treatments in RBD has increased. The research diagnostic criteria proposed for ICD-10 give an inappropriate weight to the definition used in the epidemiological studies which because of their retrospective nature are unable to observe the exact time course and clinical presentation of episodes without the bias and distortion of memory effects.

The Zurich epidemiological definition (Angst *et al.* 1988) which requires that episodes should occur at least monthly, would exclude 25 per cent of patients in the London series (Montgomery *et al.* 1989) who undoubtedly suffered from disabling RBD but who, from time to time, developed intervals longer than 1 month. Subsequent clinical studies in similar and in different populations have found the same erratic length of interval. The clinical description in ICD-10 phrases the requirement more flexibly 'occurring about monthly', but in drawing up recommendations for studies this definition may also be inappropriate.

The point of requiring a certain number of episodes in a year is not well made. Diagnostic criteria for other intermittent disorders which are less dangerous than brief depressions, for example migraine, do not require a certain frequency before the disorder is recognized although it is clear that increased frequency is associated with greater morbidity. The more important point however is that a 1-year period of illness is too long in practical terms for reliable memory and in clinical terms it is too long a period to delay treatment. In a person who develops a disabling series of brief depressions, three episodes in a period of three months should suffice as a criterion for treatment.

PHARMACOTHERAPY

The studies on the efficacy of drugs in RBD are small in number and do not provide a body of evidence from which to generalize with any confidence about the treatment of this disabling condition. The condition is difficult to study since the episodes are too brief and variable in duration for acute studies to be feasible. Likewise open reports of efficacy over periods of up to 3 months may be affected by the spontaneous remission for variable periods of between 1 and 6 months seen in some 25 per cent of our clinical sample. The only appropriate methodology to study the efficacy of a drug is to adopt a randomized group comparison against placebo using a prophylactic design. These studies are very time consuming and expensive on resources to conduct and major funding agencies have as yet been reluctant to investigate the condition.

A large single-centre placebo controlled study of the efficacy of fluoxetine in RBD in a group of recurrent suicide attempters reported that fluoxetine was without any perceptible effect on either the severity, the duration, the recurrence rates, or the suicide attempt rate in a group of 107 patients. 159 episodes on fluoxetine compared with 157 on placebo in a six month prophylactic design (Montgomery *et al.* 1992). Although this was an imperfect design for testing efficacy, for example patients who attempted suicide were withdrawn from the study, the lack of any hint of efficacy suggests that fluoxetine is unlikely to be beneficial as a treatment strategy. The role of other SSRIs as treatments for RBD remains to be tested.

There is some evidence that tricyclic antidepressants such as amitriptyline are ineffective in RBD and may even make matters worse with reports of prolonged episodes and increased aggression. However, their efficacy has not formally been tested. The evidence, such as it is, suggests that RBD has a different pharmacology to major depression not associated with apparent efficacy in major depression.

In the early studies on recurrent suicidal behaviour where brief episodes of depression of less than 2 weeks predicted subsequent suicidal behaviour on placebo, there was evidence that flupenthixol significantly reduced suicidal behaviour compared with placebo in a 6-month prophylactic design (Montgomery *et al.* 1979). This suggests that low dose flupenthixol was having some beneficial effect on the RBDs in this patient sample although this was not specifically tested. This remains the most likely pharmacological strategy for treating brief depression although the suicidal attempts may not necessarily be linked directly to the brief depressions.

There has been some interest in the efficacy of other agents in RBD based on the kindling model. This approach is based on the assumption that these brief depressions are linked to rapid cycling bipolar disorders which does not appear well founded. Preliminary investigations are negative with carbemazepine having no effect on the depressive symptoms compared with placebo in two small studies in a mainly borderline population (Cowdry and Gardner 1988; de la Fuente, personal communication).

PSYCHOLOGICAL TREATMENT

In the absence of established effective pharmacological agents, the most likely strategy for helping sufferers is to strengthen coping mechanisms. Direct intrusive psychoanalytic or behavioural techniques are often experienced as confrontational and counter-productive. Sufferers report some success with supportive, non-intrusive advice which is aimed at avoiding confrontation during the episode of brief depression. Sufferers report an increased sensitivity to criticism during the depression so important meetings or emotionally charged situations should be delayed, if at all possible, until the episode has passed. Recognizing the commonness of the disorder and the brevity of the individual episodes does much to allow the sufferer to develop strategies for coping.

CONCLUSION

The episodes of brief depression do not appear to be related to life events so that the parallels with major depression do not extend to the precipitating events seen in the majority of major depression. There may be a seasonal concentration of brief depressive episodes in some patients but this is not clearcut (Kaspar *et al.* 1992). The irregularity and unpredictable onset of the episodes which is a striking feature of the condition makes it unlikely that any of the more predictable biological cycles are involved. The episodes do not appear to be associated with menstrual periods for example. The data are suggestive rather of a disregulation and this line of investigation might be more productive. It appears as if the irregularity of the onset of the brief depression is accompanied by a more predictable correcting mechanism which keeps the episode short. An understanding of the biological mechanisms involved in the sudden onset and early resolution of these brief depressions is likely to hold the clue to the nature of depression.

REFERENCES

Angst, J. (1990). Recurrent brief depression: a new concept of depression. *Pharmacopsychiatry*, **23**, 63–6.

Angst, J. and Dobler-Mikola, A. (1985). The Zurich study: a prospective epidemiological study of depressive, neurotic and psychomatic syndromes. IV Recurrent and non-recurrent brief depression. *European Archives of Psychiatry and Neurological Science*, **234**, 408–16.

Angst, J., Vollrath, M., and Koch, R. (1988). Lofepramine in the treatment of depressive disorders. In *New aspects on epidemiology of depression* (ed. J. Angst and B. Woggon), pp. 1–14. Vieweg, Braunschweig.

Angst, J., Merinkangas, K., and Scheidegger, P. (1990). Recurrent brief depression: a new subtype of affective disorder. *Journal of Affective Disorders*, **19**, 87–98.

APA (American Psychiatric Association) (1980). *Diagnostic and statistical manual of mental disorders (DSM-III)*. APA, Washington.

APA (American Psychiatric Association) (1987). *Diagnostic and statistical manual of mental disorders (DSM-III-R)*. APA, Washington.

Bleuler, E. (1924). *Textbook of psychiatry*. Macmillan, New York.

Brown, G.W. and Harris, T. (1982). Fall-off in the reporting of life events. *Social Psychiatry*, **17**, 23–8.

Cowdry, R. and Gardner, D. (1988). Pharmacotherapy of borderline personality disorder. Alprazolam, carbamazapine, trifluoperazine and tranylcypramine. *Archives of General Psychiatry*, **5**, 111–19.

Feighner, J.P., Robins, E., Guze, S.B. *et al.* (1972). Diagnostic criteria for use in psychiatric research. *Archives of General Psychiatry*, **26**, 57–63.

Gregory, M.S. (1915). Transient attacks of manic-depressive insanity. *Medical Records*, **88**, 1040.

Jenkins, C.D., Hurst, M.W., and Rose, R.M. (1979). Life changes. Do people really remember? *Archives of General Psychiatry*, **36**, 379–84.

Kasper, S., Ruhrmann, S., Haase, T., and Moller, H.J. (1992). Recurrent brief depression and its relationship to seasonal affective disorder. *European Archives of Psychiatry and Neurological Science*, **242**, 20–6.

Kraepelin, E. (1921). *Textbook of psychiatry*. Livingstone, Edinburgh.

Lecrubier, Y. (1992). Problems in self rating recurrent brief depression. *Clinical Neuropsychopharmacology*, **15** (Suppl. 1), 9–10.

Montgomery, S.A. and Asberg, M. (1970). A new depression scale designed to be sensitive to change. *British Journal of Psychiatry*, **134**, 382–9.

Montgtomery, S.A. and Montgomery, D. (1982). Pharmacological prevention of suicidal behaviour. *Journal of Affective Disorders*, **4**, 291–8.

Montgomery, S.A. and Montgomery, D.B. (1992). Prophylactic treatment in recurrent unipolar depression. In *Long term treatment of depression* (ed. S. A. Montgomery and F. Rouillon), pp. 53–79. Wiley, Chichester.

Montgomery S.A., Montgomery, D., Rani, J. *et al.* (1979). Maintenance therapy in repeat suicidal behaviour: a placebo controlled trial. In *Proceedings 10th international congress for suicide prevention and crisis intervention*, Ottawa, pp. 227–9.

Montgomery, S.A., Roy, D., and Montgomery, D.B. (1983). The prevention of recurrent suicidal acts. *British Journal of Clinical Pharmacology*. **15**, 183S–8S.

Montgomery, S.A., Montgomery, D.B., Green, M., and Baldwin, D. (1989) Intermittent 3-day depressions and suicidal behaviour. *Neuropsychobiology*, **22**, 128–34.

Montgomery, S.A., Montgomery, D.B., Baldwin, D., and Green, M. (1990). The duration, nature and recurrence rate of brief depressions. *Neuro-Psychopharmacology and Biological Psychiatry*, **14**, 729–35.

Montgomery, D.B., Green, M., Bullock, T., Baldwin, D., and Montgomery, S.A. (1992). Has recurrent brief depression a different pharmacology? *Clinical Neuropharmacology*, **15** (Suppl. 1), 13a–14a.

Paskind, H.A. (1929). Brief attacks of manic-depressive depression. *Archives of Neurological Psychiatry*, **22**, 123–34.

Philip, M. and Benkert, O. (1992). Brief depression differentiation from minor depression. *Clinical Neuropsychopharmacology*, **15** (Suppl. 1), 17–19.

Spitzer, R., Endicott, J., and Robens, E. (1978). Research diagnostic criteria. Rationale and reliability. *Archives of General Psychiatry*, **3**, 773–82.

Vollrath, M. and Angst, J. (1989). Results of the Zurich cohort study: course of anxiety and depression. *Psychiatry Psychobiology* **4**, 307–13.

WHO (World Health Organization) (1992). *International classification of diseases, 10th revision (ICD-10)*. WHO, Geneva.

10

Mechanisms underlying recurrence and cycle acceleration in affective disorders: implications for long-term treatment

ROBERT M. POST, MARK S. GEORGE, TERENCE A. KETTER, KIRK DENICOFF, GABRIELE S. LEVERICH, and KIRSTIN MIKALAUSKAS

INTRODUCTION

The vast majority of affective disorders are recurrent (Goodwin and Jamison 1984). Some 70 per cent of patients with a first episode of unipolar affective illness will have another episode and virtually all patients with bipolar illness will have multiple episodes over the course of their lives. In a substantial subgroup of these patients, evidence of cycle acceleration may occur. Kraepelin (1921) was among the first to comment on the progressively decreasing 'well intervals' between successive episodes of affective illness (Fig. 10.1). As summarized in Table 10.1 and reviewed elsewhere

TABLE 10.1. *Cycle acceleration in affective illness: studies reporting a decreasing well interval between successive episodes in bipolar disorder*

Reference	Number of subjects
Swift 1907	105
Kraepelin 1921	903
Paskind 1930	633
Lundqvist 1945	319
Angst and Weiss 1967	388
Bratfos and Haug 1968	207
Grof *et al.* 1974	987
Taschev 1974	652
Zis *et al.* 1980	334
Roy-Byrne *et al.* 1985	95
10 studies	4623

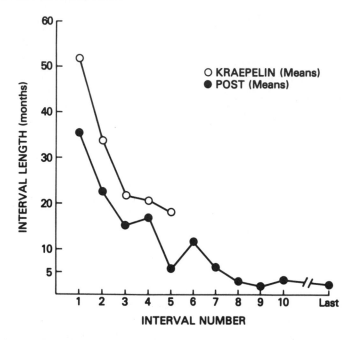

FIG. 10.1. Decreasing well intervals in recurrent affective illness.

(Cutler and Post 1982; Post *et al.* 1984), this pattern of cycle acceleration appears to be the modal one in a large number of studies in more than 5000 subjects. However, this pattern is not invariant. Some patients will demonstrate the stable course of illness and a few even appear to demonstrate a spontaneous recovery or burnout. Some patients may also begin their illness with a pattern of rapid cycling (four episodes/year) from the onset (Grof *et al.* 1974; Zis and Goodwin 1979; Roy-Byrne *et al.* 1985). However, many patients with unipolar or bipolar affective illness have relatively longer well intervals between episodes early in their illness and, with successive episodes, cycle frequency increases. This pattern is not invariably unidirectional, however, as some patients with ultrarapid cycling illness (episodes lasting days to weeks) can undergo periods of cycle deceleration either spontaneously (Cutler and Post 1982), or more particularly, following adequate treatment interventions (Post 1990a). Nonetheless, as many patients do show this pattern of cycle acceleration, it is of great theoretical as well as clinical importance. If mechanisms underlying cycle acceleration are adequately delineated, this progressively deteriorating component of the illness could be prevented and/or reversed.

In addition to cycle frequency, other indices may reflect the progressive aspects of the illness. These might include increased severity of episodes over time, more precipitous onsets to individual episodes (Post *et al.*

1981*a*), a transition from episodes that are triggered by psychosocial and other stressors to those that are occurring spontaneously (Post 1992), and the development of refractoriness to therapeutic interventions. Endogenous and exogenous mechanisms that might be associated with these phenomena as well as cycle acceleration will be briefly discussed in this manuscript. We suggest that these putative mechanisms underlying a tendency for progressive deterioration of the course of affective illness as well as its transition from precipitated to autonomous episodes speaks to the importance of early institution and long-term maintenance of pharmacoprophylaxis.

ROLE OF SOCIAL STRESSORS EARLY BUT NOT LATE IN RECURRENT AFFECTIVE DISORDERS

A robust literature implicates psychosocial stress in the aetiopathogenesis of the affective disorders (Paykel 1979, 1982; Ellicott *et al.* 1990). At the same time, neurobiological investigators emphasize the endogenicity of some depressions with full-blown episodes occurring 'out of the blue'. We suggest that the earlier observations of Emil Kraepelin (1921) and more recent systematic observations in a number of studies indicate that these may not be mutually exclusive perspectives (Post 1992). Rather, it is increasingly apparent that psychosocial stressors may play a role in the precipitation of initial episodes, but with successive recurrences, as the system is 'sensitized', stressors may play a less prominent role until finally episodes occur with relative autonomy, which often appears to be the case in patients with rapid-cycling affective disorders (Fig. 10.2).

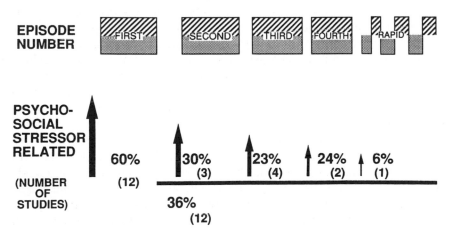

FIG. 10.2. Precipitants of affective illness: greater relationship to first compared to subsequent episodes.

As reviewed elsewhere (Post 1992), clinical observations from 25 investigators and from 12 controlled studies are all compatible with the idea that psychosocial stressors may be implicated in the first episode of major affective illness of either a manic or depressive variety approximately 60 per cent of the time, but then appear less intimately involved with second or third episodes where stressors are implicated in only 36 per cent. In addition, many of these investigators often emphasize the role of very early life events and psychosocial stresses, losses, and, particularly, early idiosyncratic life experiences pertaining to social and affiliative roles and self-esteem. When such a predisposed individual is rechallenged with a 'matching event' or recurrence of the stressor, it may then become sufficient to evoke a full-blown episode, even though the initial stresses were without this full-blown effect. In this fashion, one conceptualizes the potential role for early life experiences providing a lasting vulnerability to stressor reactivation.

KINDLING AS A MODEL FOR THE TRANSITION FROM PRECIPITATED TO SPONTANEOUS EPISODES

The kindling paradigm helps to conceptualize mechanisms involved in both the increasing responsivity to the same inducing stimulus over time, and, with sufficient repetition of the syndrome, the transition from triggered episodes to those occurring spontaneously (Post *et al.* 1986) (Table 10.2). In the development of amygdala-kindled seizures, animals receive intra-cerebral stimulation, for example, of the amygdala once a day for 1 s. Initially, no behavioural or electrophysiological effects are discerned. However, with repeated daily stimulation, the threshold for after-discharge development decreases, after-discharges begin to occur with increasing duration and complexity in the amygdala and then begin to spread throughout the neuroaxis to synaptically-related structures. This can be mapped with electrophysiological recording, metabolic mapping with deoxyglucose, or, most recently, with *in situ* hybridization for the proto-oncogene c-fos (Clark *et al.* 1991). In these latter studies, Clark and colleagues have demonstrated that the proto-oncogene c-fos is initially induced unilaterally in the piriform cortex and dentate gyrus but, with successive stimulations, becomes bilateral and involves other areas of the brain as one moves through kindled seizure stage evolution: from stage 1 seizures which involve only behavioural arrest; to stage 3 which involves the forepaws unilaterally; to stage 4 which is bilateral seizures of the trunk and forepaws; followed by stage 5 seizures which include rearing and falling as well. Thus, a process evolves with repeated stimulations that

TABLE 10.2. *Phenomena in course of affective disorders modelled by kindling and behavioural sensitization*

Descriptors	Kindling (K)	Sensitization (S)	Phenomenon
Stressor vulnerability	++	++	Initial stressors early in development may be without effect but predispose to greater reactivity upon rechallenge
Stressor precipitation	++	++	Later stress may precipitate full-blown episode
Conditioning may be involved	–	++	Stressors may become more symbolic
Episode autonomy	++	–	Initially precipitated episodes may occur spontaneously
Cross-sensitization with stimulants	–	++	Co-morbidity with drug abuse may work in both directions affective illness ⇌ drug abuse
Vulnerability to relapse	++	++	S and K demonstrate long term increases in responsivity
Episodes may: (a) become more severe	++	–	S and K both show behavioural evolution in severity or stages
(b) show more rapid onsets	++	++	Hyperactivity and stereotypy show more rapid onsets
Anatomical and biochemical substrates evolve	++	±	K memory-trace evolves from unilateral to bilateral
IEGs involved	++	++	Immediate Early Genes (IEGs) such as c-fos induced
Alterations in gene expression occur	++	++	IEGs may change gene expression, especially of peptides over long time domains
Change in synaptic microstructure occurs	++	–	Neuronal sprouting and cell loss indicate structural changes
Pharmacology differs as function or stage of evoluton	++	++	K differs as a function of stage; S differs as a function of development versus expression

yield an increasing reactivity of the brain that occurs in a long-lasting fashion and is associated with a marked facilitation of the behavioural consequences of this stimulation from no behavioural effect at all to minor changes in behaviour to full-blown seizure episodes (Goddard *et al.* 1969; Racine 1978). If these episodes are repeated a sufficient number of times, the animal develops spontaneity, that is i.e. spontaneous seizures in the absence of electrophysiological stimulation (Wada *et al.* 1974; Pinel and Rovner 1978*a,b*; Pinel 1981). That is, the kindled seizure process has been so facilitated by repetition that it undergoes a transition from requiring triggering by exogenous stimulations to a phase where it now occurs autonomously.

Clearly, kindled seizures present a highly nonhomologous model (Post *et al.* 1991*a*) for affective disorders in that few components of the behaviour in rats involved in amygdala-kindled seizures represent anything like that which occurs in the affective disorders; moreover, patients with affective disorders do not demonstrate a propensity for the development of seizures. None the less, we suggest the possibility that parallel processes in very different neurobiological systems may be engaged by psychosocial stressors which are initially capable of triggering minor dysphoric episodes and then,

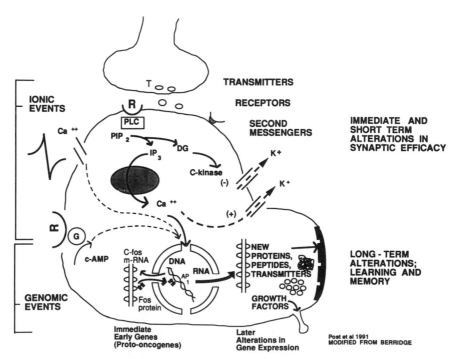

FIG. 10.3. Neuromodulation and memory.

finally, more major episodes of mania and depression. In addition, as in the kindling model, with sufficient repetitions of affective episodes these, too, undergo the transition to spontaneity and begin to emerge in the absence of critical psychosocial triggers.

The kindling model also helps us conceptualize the classes of neuro-biological mechanisms that may underlie these memory-like processes associated with the development of the kindled trace and its permanent change in responsivity for the organism (Goddard and Douglas 1975) (Table 10.2). Initial evidence suggests that not only are the proto-oncogenes such as c-fos, which are members of the immediate early gene transcription family, useful in marking neurons that have been activated (Clark *et al.* 1991), but as well may be involved in a cascade of neuro-biological events that are intimately involved in the process of laying down memory traces that may be occurring at the level of alterations in gene transcription (Berridge 1989; Crabtree 1989; Dragunow *et al.* 1989; Post *et al.* 1990*c,d*, 1992*a*; Anokhin and Rose 1991; Morgan and Curran 1991; Rose 1991). In this fashion, it is possible to conceptualize how early experiences with stressful life events as well as the experiences of recurrent affective episodes themselves could alter neurobiological processes in a long-lasting fashion leaving behind a vulnerability for recurrence (Post *et al.* 1991*b*; Post 1992).

IMPACT OF STRESSORS AND EPISODES AT LEVEL OF GENE EXPRESSION: LESSONS FROM THE KINDLING MODEL

Neurotransmitter alterations associated with ongoing neural activity not only affect short-term adaptive changes at the level of neurotransmitters or receptors, but also longer-term adaptations at the level of gene expression. While depolarized cells (i.e. those firing in the process of neurotrans-mission) are activated by second messenger systems such as calcium and cyclic-AMP, these processes not only have direct cytoplasmic effects on synaptic transmission, but they also have indirect nuclear effects in induc-ing proto-oncogenes such as c-fos, c-jun, or fos-related antigens (fras) as well as a variety of other transcription factors (Fig. 10.3). C-fos, c-jun, and jun-B are members of the family of oncogenes, called immediate early genes (IEG), which appear in a period of minutes to hours but then may initiate effects with longer lasting time domains. In the case of the mRNA for c-fos, this then initiates a process of construction of the Fos protein on the endoplasmic reticulum of the cell and Fos protein is transported into the nucleus where, linked by leucine zippers to other proto-oncogenes such as Jun, it forms heterodimers that bind to AP1 sites on DNA, which are important in initiating gene transcription. Other Fras are formed with different time domains and these may bind with Fos or Jun and be involved either in the enhancement or repression of transcriptional activity. Jun-B also appears highly involved in negative transcriptional control. Moreover, there are many different types of transcription factor with different DNA binding motifs (leucine zipper, zinc fingers, copper fists (Vinson *et al.* 1989; Shuman *et al.* 1990; Millbrandt 1991) and these may even demonstrate interactions or cross-talk. For example, c-fos effects may be inhibited by glucocorticoid receptor effects and vice versa (Lucibello *et al.* 1990; Yang-Yen *et al.* 1990).

While the induction of c-fos has not been definitively linked to longer term consequences for the cell, it is highly likely that the induction of mRNA for and synthesis of preproenkephalin (Sonnenberg *et al.* 1989) and other neuropeptides such as CRH, TRH, somatostatin and dynorphin are influenced in this cascade (Kubek *et al.* 1990; Smith *et al.* 1991; Rosen *et al.* 1993). These latter changes in neuropeptides can last for days to weeks, or even longer (Kato *et al.* 1983), and provide one set of mechanisms by which acute changes in neurotransmitter activity could be transduced into other biological changes with a time frame relevant to the recurrent affective disorders. This model may also be pertinent to the long-lasting effects of stressors as a variety of stressors, have also been shown to induce c-fos (Daval *et al.* 1989; Nakajima *et al.* 1989*a*,*b*). The kindling model also demonstrates that longer lasting changes in neuronal and synaptic microsynaptic structure may occur, with increases in both sprouting and cell death having been documented (Cavazos and Sutula 1990; Sutula 1990).

It is also possible that these types of biochemical and microstructural changes could underlie long-lasting alterations in behavioural and synaptic responsivity relevant to the course of recurrent affective disorders; i.e. both stressors and episodes themselves may leave behind memory traces (Post 1992). Microstructural changes have been reported not only in kindling, but also with models of learning and memory, including long-term potentiation (LTP) and single-trial passive-avoidance learning in the 1-day old chick (Anokhin and Rose 1991; Rose 1991). In a series of elegant studies by Steven Rose, they demonstrated that a chick trained to peck at the shiny bright object would permanently avoid that object if it was associated with a bitter tasting solution. This type of learning was associated not only with the early induction of the proto-oncogenes c-fos and c-jun, but, in a longer time-frame, was associated with changes in the spine and synaptic density as well. The anatomical localization of the memory trace also changes over time in a process similar to that observed in kindling (Clark *et al.* 1991) and in human memory formation (Mishkin and Appenzeller 1987; Squire and Zola-Morgan 1991). Whether or not similar types of structural alterations are ultimately demonstrated to occur in the affective disorders, we would suggest the possibility that long-term biochemical if not structural adaptations do occur in response to stressors and the occurrence of repeated episodes in a fashion that leads to the altered vulnerability for recurrence. This vulnerability may be manifested in terms of cycle acceleration, more precipitous onsets of individual episodes, as well as the transition from precipitated to autonomous episodes.

Elsewhere, we have discussed how the model of behavioural sensitization to the psychomotor stimulant cocaine may also be a useful one for the progressive evolution of manic episodes; in particular, dysphoric mania (Post and Weiss 1989; Post *et al.* 1989, 1991*a*) (Table 10.2). In the case of sensitization, biochemical mechanisms may be different from those involved in kindling, but also involve the induction of the proto-oncogene c-fos through a D1 receptor mechanism (H.A. Robertson *et al.* 1989; Young *et al.* 1989, 1991; Graybiel *et al.* 1990; G.S. Robertson *et al.* 1990). Additionally, we have documented a prominent context-dependent or conditioned component of behavioural sensitization where animals appear more reactive to each dose of the drug if the environmental context cues are similar to their initial experience, but not if equal doses of cocaine are given in a different environment where the conditioned cues are not available (Post *et al.* 1981*b*, 1987; Weiss *et al.* 1989). Pert and associates (Fontana *et al.* 1991, 1993) have recently documented increases in dopamine overflow in this context-dependent sensitization model, suggesting that important neurochemical events occur in association with conditioning. While the detailed mechanisms underlying the long-lasting vulnerability to increase behavioural responsivity in the sensitization model has not all been definitively elaborated, these data suggest that important mechanisms at the level of both transmitter release and, possibly, at the level of gene transcription (based on the induction of c-fos) could be involved in this process which provides a relatively homologous model for the recurrence of manic episodes.

DRUG-INDUCED SWITCHES, CYCLE ACCELERATION, AND PROGRESSION TO SPONTANEITY

Pickar *et al.* (1982) have suggested that sensitization may occur to manic episodes triggered by psychopharmacological agents. They observed, in patients who switched into mania on monoamine oxidase inhibitors (MAOIs) during the first clinical trial, that the duration time to the switch on a second exposure to the MAOIs in a double-blind trial was consistently shorter than that initially observed. Some patients may demonstrate only antidepressant-induced switches but, in other patients, antidepressants are associated with the first induction of a manic episode and then these manic episodes may begin to occur spontaneously in the absence of the anti-depressant triggers. These data suggest the possibility that, as in the case of stressors reviewed above, there may be a transition from triggered episodes to those occurring on a more autonomous basis.

In addition to the precipitation of single episodes of mania, tricyclic antidepressants are also implicated in cycle acceleration (Kukopulos *et al.* 1980, 1983; Wehr and Goodwin 1987*a,b*; Rouillon 1991). Wehr and Good-win (1987*a,b*) have reviewed the data suggesting that in a subgroup of vulnerable patients, application of tricyclic antidepressants may be associ-ated with an increased frequency of cycling which then is attenuated on drug withdrawal. In several instances they were able to establish this finding in an on–off–on fashion, convincingly demonstrating that these antidepressant agents may shorten depressive episodes at the cost of short-ening the entire cycle duration and the time to the next recurrence in a subgroup of vulnerable individuals.

In our series of subjects studied at the NIMH, we have attempted to classify those patients who had likely or definite precipitation of manic episodes and definite or likely cycle acceleration during treatment with these agents (L.L. Altshuler, R.M. Post, G. Leverich, K. Mikalauskas, Rosoff, and L. Ackerman, unpublished manuscript). We found that approximately 35 per cent of our bipolar patients experienced antidepres-sant-induced episodes and approximately 25 per cent demonstrated clear-cut evidence of cycle acceleration. The tricyclic-induced 'switches' had some evidence of a more accelerated course of illness as evidenced by a trend towards increased numbers of episodes in the year prior to NIMH admission. The nine patients with tricyclic- or MAOI-induced cycle accele-ration showed an earlier age of onset and a greater number of episodes in the year prior to NIMH hospitalization (10 ± 6.3 versus 4.9 ± 4.4, $p<0.02$), although they had significantly fewer hospitalizations prior to NIMH admission compared with 26 patients without such clear-cut evidence.

Our data in patients who are largely initially classified as rapid cycling patients may help clarify how some of the discrepancies in the literature between those who indicate there is a high incidence of antidepressant-induced switch (Bunney *et al.* 1970; Kukopulos *et al.* 1980, 1983; Wehr and Goodwin 1987*a,b*; Wehr *et al.* 1988; Himmelhoch *et al.* 1991) compared with those where this does not appear to be a major problem (Lewis and Winokur 1982, 1989; Angst 1986; Kupfer *et al.* 1988). Most of the groups reporting an important incidence of tricyclic-induced switch in cycle acceleration appear to be dealing with patients who have largely refractory and/or rapid-cycling disorders. Groups not reporting a substantial incidence of these phenomena may be dealing with a different population characterized by a more intermittent course of illness and slower initial cycle frequency to begin with. However, a recent large prospective study of Rouillon *et al.* (1989, 1991) suggests that the heterocyclic maprotiline, even when used in small doses, may be associated with a small but significant increase in the switch rate compared with placebo seen in unipolar patients. The controversy regarding tricyclic-induced switches clearly requires further investigation as adjunctive antidepressants are often used in the treatment of breakthrough bipolar depressions. A recent article by Himmelhoch *et al.* (1991) suggested that the MAOI tranylcypromine was markedly more effective an antidepressant than the tricyclic antidepressant imipramine in anergic bipolar depressions without increasing the probability of a switch over that observed with the tricyclics. None the less, the MAOIs do not appear to be without the liability of cycle induction in one series (L.L. Altshuler, R.M. Post, G. Leverich, K. Mikalauskas, Rosoff, and L. Ackerman, unpublished manuscript), although the ability to decrease cycle frequency has been reported for the selective monamine oxidase A inhibitor clorgyline (Potter *et al.* 1982).

Kukopulos and others (1980, 1983) have raised the issue that use of tricyclics may be associated with not only the induction of manic episodes, but also the possibility of cycle acceleration and conversion to a continuous cycling pattern. Adequate prospective studies are needed to further elucidate the methodological soundness of these observations. However, clinicians are left with critical decisions regarding the use of these agents in the long-term treatment of bipolar depressive patients, particularly those who have episodes that break through prophylaxis with lithium, carbamazepine, or valproate. Although preliminary evidence suggested that bupropion might be less likely to induce mania and cycling (Shopsin 1983; Haykal and Akiskal 1990), two recent case reports of mania in patients receiving this medication (Bittman 1991; Zubieta 1991) make evident the need for further information.

EVOLUTION TO ULTRA-ULTRA RAPID CYCLING AND THE IMPACT OF TREATMENT: IMPLICATIONS OF CHAOS THEORY

Until the last several years, cycling at frequencies of 48 h were assumed to be the maximum degree of cycle acceleration achieved in bipolar affective disorders. However, there is increasing recognition that some patients, even with classical illness, can demonstrate distinct oscillations in mood at frequencies exceeding this every other day pattern. Kramlinger and Post (1988) highlighted these occurrences in a small series of patients, suggesting that a continuum of cycle frequency existed from those with nonrapid cycling to rapid cycling (>4 episodes/year) to ultra-rapid cycling (regular occurrences of episodes with a pattern of days to weeks) and, finally, ultra-ultra rapid (ultradian) cycling where mood fluctuations occurred on the

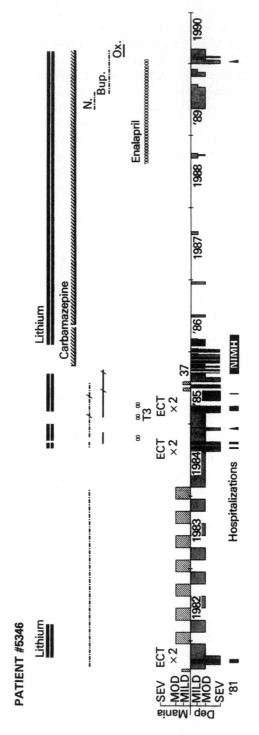

FIG. 10.4. Loss of prophylactic efficacy in a rapid-cycling bipolar II female patient.

order of minutes to hours, with several episodes occurring in a single day.

M.S. George, M. Jones, R.M. Post, F. Putnam, K. Mikalauskas, and G. Leverich (unpublished manuscript) have further elaborated on this concept following the observation that many of these patients observed in our hospital arrived at this pattern of ultra-ultra rapid cycling from slower cycle frequencies at their outset of illness which progressively accelerated with both time and repetition. Such an example is illustrated in Fig. 10.4. The patient (No. 5346) demonstrated the onset of bipolar affective disorder with compound episodes in 1973 of hypomania followed by depression with another triphasic series of episodes in 1976 and a depression in 1978. Each of these was followed by an approximately 2-year well interval. In 1981, a mild hypomanic episode was followed by a more severe depression which resulted in hospitalization, the administration of several electroconvulsive treatments, and the institution of antidepressant treatment. With this treatment and in spite of lithium augmentation, the patient demonstrated a more rapid fluctuation in mood with continuous cycling between mania and depression. Following the discontinuation of antidepressant treatment in 1984, a more prolonged depressive period occurred which resulted in a series of hospitalizations and ECT treatment in the context of various pharmacotherapeutic manipulations including antidepressants and several trials with lithium, all of which were ineffective in slowing down her rapid oscillations between severe depression and either mild depression, euthymia, or hypomania. This pattern was observed during a period of double-blind placebo substitution at the NIMH and discrete periods of euthymia were observed to punctuate more severe periods of immobilizing and incapacitating depressions lasting between 1 and 7 days. However, when the patient and nursing staff employed two-hourly ratings, periodic sharp fluctuations in mood from severe depression to euthymia and even mild hypomania were observed to occur at frequencies faster than 24 h. Thus, this patient's pattern of illness appeared to show both progressive cycle acceleration and, at the faster periodicities, fluctuations that were chaotic. We mean chaotic in the mathematical sense of the term implying nonlinear variation that obeys certain rules and may not necessarily be truly random. As in the principles of fractal geometry, where self-similarity of patterns are observed across successive temporal or macro- to microscopic dimensions, similar patterns of mood fluctuations could be observed on a time dimension of years, days, or hours depending on the intensity of temporal rating scales employed (Kramlinger and Post 1988).

The patient in Fig. 10.4 went on to have a partial response to treatment with carbamazepine and a complete cessation of episodes when lithium (which had previously been ineffective used either alone or in combination with tricyclic antidepressants) was added to the carbamazepine treatment. The patient remained well for 3 years and then began to demonstrate the

re-emergence of progressively more severe depressive episodes in a tolerance pattern that we have previously observed in a subgroup of patients
with successful long-term response to carbamazepine (Post 1990*a*; Post *et
al.* 1990*b*). This pattern of tolerance development is of interest in its own
right and has important treatment implications (shifting to other drugs that
do not show cross tolerance (Weiss and Post 1991) or discontinuing and
restarting the same agent (Pazzaglia and Post 1992). However, in the
context of the present discussion, it is of particular interest from the
perspective that a subgroup of patients may show not only cycle acceleration, but evidence of a malignantly progressive underlying process,
evidenced also by re-emergence of episodes breaking through initially
successful pharmacoprophylaxis.

This patient's case history is also characteristic of others who have shown
a pattern of cycle acceleration and conversion from intermittent to continuous episodes under the influence of antidepressant treatment. George
et al. (1993) have used mathematical modelling and chaos theory to suggest
the possibility that progressive cycle acceleration such as demonstrated by
this patient may be attributable to slow increases in a single variable in a
mathematical formula that has been used in other systems demonstrating
chaos. Bauer and Whybrow (1991) have presented a function, $x[n] = r
\cdot x[n-1] \cdot (1 - x[n-1])$, which has interesting properties similar to the
mood oscillations seen in patients with bipolar disorders. The iterative
nature of this function $x[n]$, with later values of the function being dependent on earlier values, is the mathematical analogue of our hypothesis that
earlier affective episodes beget later ones. Changes in the parameter r can
yield dramatic changes in the behaviour of $x[n]$. With increasing r the
function $x[n]$ displays increasing instability, periodicity, and finally, chaotic oscillation. This mathematical model shows that given an iterative
process (later affective episodes contingent on earlier affective episodes),
changes in a single biological parameter (r) can yield dramatically different
longitudinal causes.

Thus, small variations in a single factor can change the pattern from
isolated intermittent episodes ($r = 1$) to more continuous rapid fluctuations
($1 > r < 0.25$) and, finally, very fast frequencies of oscillations with a
chaotic pattern ($3 > r < 4$) (M.S. George, M. Jones, R.M. Post, F.
Putnam, K. Mikalauskas and G. Leach, unpublished manuscript). The
patient illustrated in Fig. 10.4 shows these transitions: intermittent episodes, 1973–81; continuous, regular, rapid cycles, 1982–3; ultrafast chaotic
oscillations, 1985. While the ultimate utility and validity of this mathematical equation (or others) for modelling the progressive course of affective
disorders remains to be directly demonstrated, it nevertheless demonstrates the possibility that alterations in a single biological factor (in this
model represented by 'r') could underlie progressive enhancements in

pattern and cycle frequency observed in the course of the affective disorders. It also leaves the possibility that pharmacotherapy and treatment response may vary markedly as a function of the illness (Post *et al*. 1986). That is, drugs effective in the period when $1 > r < 0.25$ may prove ineffective at later stages, or vice versa. This system has very different properties at different stages (much as the behaviour of air over a wing varies with airspeed, and functionally changes at low speed, causing turbulence with stalling).

Whatever this biological variable turns out to be, we suggest that in bipolar patients it is potentially adversely impacted by stressors and tricyclic and MAOI antidepressants and can be attenuated by lithium and other anticycling agents. The anticonvulsants carbamazepine and valproate which have recently been demonstrated as playing an important role in some lithium-refractory manic-depressive patients (Post 1990*a,b*), also would appear to be able to positively impact on this biological variable underlying cyclicity and cycle accleration in the affective disorders. We are not suggesting that the tricyclic or MAOI antidepressants are the only cause of cycle acceleration, as this pattern of progressive deterioration and the frequency and pattern of illness had been observed in the pre-psychopharmacological era by Kraepelin and many other observers and investigators (see reviews of Cutler and Post 1982; Squillace *et al*. 1984; Post *et al* 1986). What we are suggesting is that in this general pattern of progressive deterioration in the illness, tricyclic antidepressants can further accelerate cycle frequency and, as suggested by Kukopulos and the current case demonstration, convert the patient from an intermittent to a continuous form of the illness.

Since lithium, carbamazepine, and valproate each appears capable not only of inhibiting acute manic episodes but also preventing the new occurrence of both manic and depressive episodes, it would appear that these agents are all able to interfere with the biological parameters involved in generating acute episodes as well as their subsequent recurrence and accleration over time, as portrayed in the chaos model. While the mechanism of action of carbamazepine, valproate, and lithium has not been definitively demonstrated, numerous biochemical candidate systems have been identified for each of these agents (Bunney and Garland–Bunney 1987; Post 1987, 1988; Post *et al*. 1992*b*). Of interest in the current analysis, alterations in a variety of second messenger systems involving cyclic-AMP and calcium have not only been postulated to occur in the affective disorders, but also in the mechanism of action of these agents. Moreover, oscillations in these variables have recently been postulated to underlie disorders of cyclic cell division, circadian rhythms, and other oscillatory phenomena. Taken together, these pathophysiological and therapeutic perspectives suggest the possibility (notwithstanding the complexity of the biology of the mood disorders) that alterations in and amelioration of single biological processes could underlie both episode appearance and cycle acceleration, as well as their treatment with therapeutic agents.

LITHIUM DISCONTINUATION-INDUCED REFRACTORINESS

The development of treatment-refractoriness can be another manifestation of the deteriorating course of some mood disordered patients. Some patients develop refractoriness spontaneously or present with it either initially or in the course of drug treatment (Post *et al.* 1990*b*). We have also observed a small series of patients in whom long periods of successful and complete lithium prophylaxis was terminated by the recurrence of an episode following discontinuation of lithium treatment (Post 1990*a*; Post *et al.* 1992*c*). Tragically, upon reinstitution of lithium therapy in these patients, a therapeutic response was not again achieved. Numerous other investigators have informed us that they have also seen this occurrence and are now advising conservative approaches to patients wishing to discontinue long-term successful lithium prophylaxis. These data are raised in the context of the current discussion, as they suggest that not only may patients demonstrate a pattern of increasing cycle frequency over time, implying that episodes facilitate the occurrence of further episodes, but the recurrence of a new episode may ultimately also affect pharmacotherapeutic response.

Should this empirical observation of lithium discontinuation-induced refractoriness be documented in a large series of patients, it would add further support to the notion that episodes sensitize to further episodes and after a sufficient number of episodes, even in previously excellent responders to lithium, nonresponsiveness may supervene. These data are, at least in part, consistent with the recent data of Gelenberg and associates (1989), indicating that lithium (either in high or low dosage forms) is not as effective in patients who had more than three episodes compared with those with fewer prior episodes. These data, gleaned in a systematic prospective clinical study support numerous previous series indicating that lithium is less effective in rapid cycling patients compared with those with fewer prior cycles (see review of Post *et al.* (1990*a*)).

O'Connell *et al.* (1991) also report that the best predictor of lithium-nonresponsives was greater numbers of episodes prior to study entry: 3.8 episodes in good responders, 5.3 in fair, and 7.7 in poor. If it is possible to engender lithium refractoriness based on discontinuation of effective treatment, one might also raise the possibility that repeated episodes of noncompliance could also engender refractoriness in the otherwise responsive patient.

SENSITIZATION AND KINDLING PERSPECTIVES:
IMPLICATIONS FOR LONG-TERM PROPHYLAXIS IN
UNIPOLAR AND BIPOLAR DEPRESSION

Observations of discontinuation-induced refractoriness in bipolar patients
also raise the question of whether similar phenomena occur in the treatment of unipolar affective disorder. While there is a wide consensus of the
success of a variety of prophylactic strategies in the long-term treatment of
unipolar depression with drugs, including nonselective antidepressants
(Prien *et al.* 1973; Coppen *et al.* 1978; Kane *et al.* 1982; Glen *et al.* 1984;
Frank *et al.* 1990), MAOIs (Georgotas *et al.* 1989; Robinson *et al.*
1991), serotonin-selective antidepressants (Bjork 1983; Glen *et al.* 1984;
Montgomery *et al.* 1988; Jakovijevic and Mewett 1991; Montgomery and
Dunbar 1991; Doogan and Caillard 1992), noradrenergic-selective antidepressants (Rouillon *et al.* 1989), and even lithium (Prien *et al.* 1973,
1984; Schou 1979; Kane *et al.* 1982; Glen *et al.* 1984; Souza *et al.* 1990)
there appears to be a general under-appreciation and under-utilization of
long-term prophylaxis in unipolar illness, particularly in the USA. That is,
conventional practice, even with the patient who has had a series of
depressive episodes in the past several years, has been to treat the patient's
acute episode and engage in continuation therapy of some 4–6 months'
duration. At that point, drug treatment is often discontinued at the suggestion
of the treating physician or at the wish of the patient, without systematic
review of the overwhelming evidence for the benefits of prophylaxis. This
perspective is often supported by the general popular view that depressive
illness is a mild disorder (rather than potentially disabling and lethal), all in
the patient's head (i.e. a mental phenomenon with biological basis), can be
successfully countered with will or effort (it's your fault), will run its course
(spontaneously cure itself forever), and does not require major, long-
term treatment with drugs (which should be discontinued as soon as possible because they are either 'bad' like drugs of abuse or potentially
toxic).

However, the empirical data on recurrence and the sensitization and
kindling perspective we have elaborated above (Table 10.2), as well as the
possibility that discontinuation of treatment with the emergence of new
episodes could engender refractoriness as documented in a small series of
patients during lithium treatment, raise the possibility that this pattern of
intermittent treatment could have potentially adverse consequences, not
only on recurrences, but on the course of unipolar illness as well. In
successfully treated unipolar depressive patients with several prior recurrences, placebo substitution results in a depressive relapse in some 50 per
cent of subjects by one year (Prien *et al.* 1973, 1984; Kane *et al.* 1982; Bjork

TABLE 10.3A. *Prophylaxis of unipolar depression: controlled studies*

	% Relapse		
	Placebo	Rx	Reference
Non-selective			
Amitriptyline (1)	31	0*	Coppen *et al.* 1978
Amitriptyline (3)	88	43†	Glen *et al.* 1984
Imipramine (1)	100	67	Kane *et al.* 1982
Imipramine (1)	23	12†	Jakovijevic and Mewett 1991
Imipramine (2)	85	29*	Prien *et al.* 1973
Imipramine (2)	71	44†	Prien *et al.* 1984
Imipramine (3)	78	21‡	Frank *et al.* 1990
NE selective			
Maprotiline (1)	32	16*	Rouillon *et al.* 1989
5-HT selective			
Zimeldine (1½)	84	32‡	Bjork 1983
Sertraline (1)	46	13‡	Doogan and Caillard 1992
Fluoxetine (1)	57	26‡	Montgomery *et al.* 1988
Paroxetine (1)	39	15*	Montgomery and Dunbar 1991 Jakovijevic and Mewett 1991
Paroxetine (1)	23	14	Jakovijevic and Mewett 1991
5-HT$_{1A}$			
Buspirone ( )		†	Fabre 1991
Lithium			
Lithium (1)	84	29‡	Schou 1979
Lithium (1)	100	29‡	Kane *et al.* 1982
Lithium (2)	85	41†	Prien *et al.* 1973
Lithium (2)	71	57†	Prien *et al.* 1984
Lithium (2)	58	8†	Souza *et al.* 1990
Lithium (3)	88	42†	Glen *et al.* 1984
MAOI			
Phenylzine (1)	65	13†	Georgotas *et al.* 1989
Phenylzine (2)	75	10‡	Robinson *et al.* 1991

* $p < 0.01$.
† $p < 0.05$.
‡ $p < 0.001$.
( ) Indicates probability of recurrence.

1983; Glen *et al.* 1984; Montgomery *et al.* 1988; Frank *et al.* 1990; Souza *et al.* 1990) (Table 10.3A). When these patients, typically with about three episodes of unipolar illness in the prior 2–3 years were maintained on long-

TABLE 10.3B. *Impact of prophylaxis on relapse rates in unipolar depression*

	Placebo (%)	Active (%)
1-year trials	55	21
(*N* = 11 studies)		
2-year trials	74	32
(*N* = 6 studies)		
3-year trials	85	35
(*N* = 3 studies)		
All trials	65	26
(*N* = 20)		

term prophylaxis, the literature suggests that there is at least a 50 per cent lesser likelihood of relapse at 1, 2 and 3 years compared with placebo substitution (Table 10.3B). Thus, the potential morbidity and mortality associated with these recurrences could be avoided. Statistics on rate of first recurrence in patients on placebo versus extended prophylaxis may also underestimate the potential morbidity, as more than one episode may occur over a period of time in this untreated phase. In addition, we are suggesting the possibility that, following the occurrence of new episodes, patterns of chronicity and refractoriness to treatment could also potentially be engendered by repeated discontinuation of an effective treatment, a proposition that deserves to be systematically examined in both retrospective and prospective data sets.

Several reviews have recently documented the high rate of recurrence of new episodes following lithium discontinuation. If lithium is discontinued suddenly, some 80–90 per cent of patients relapse. However, if lithium is discontinued with a tapering, there may be some lessening of the rate of relapse and lengthening of the time to the first recurrence (Suppes *et al.* 1991; Faedda *et al.* 1993). Given these data and the additional potential risk of discontinuation-induced refractoriness, the current perspective suggests the utility of a highly conservative approach to the long-term maintenance of prophylaxis in bipolar disorder. In addition, given the highly recurrent nature of the illness, there is increasing appreciation of the importance of early institution of prophylaxis. For example, Strober *et al.*

(1990) have observed a 92 per cent relapse rate within the first 18 months of discontinuing treatment in adolescent manic patients. This compares with some 37.5 per cent of patients who relapse in this 18 month period if they continue on their lithium. Thus, in the bipolar illnesses, particularly in those patients with a family history of affective disorder, institution, and maintenance of pharmacoprophylaxis after the first major episode should be considered and almost all of the authorities would agree that this should be performed after the second manic episode occurring within a relatively short-time frame.

However, the field is in a lesser degree of consensus around the issue of institution and maintenance of prophylaxis in unipolar affective illness. Given the consistency of reports indicating that a variety of agents are capable of reducing recurrences by greater than one-half at 1-, 2-, or 3-year time-frames of study, perhaps more aggressive use of prophylaxis should now be considered for unipolar illness as well. Many investigators would agree that after three major episodes of depression in a relatively short period of time (several years), pharmacoprophylaxis should be instituted. It would appear that initiating prophylaxis after the first episode of unipolar depression would treat a large number of patients who are not 'destined' to have another episode. The recent data of Angst and associates (1990) suggest that some 35 per cent of patients with unipolar affective disorders in a community sample were demonstrated to have the recurrent variety. Thus, it becomes important to identify and follow these patients and target them for early prophylaxis if two or three episodes have occurred in the past several years.

Given the very high rate of recurrence alone in this subgroup when they are placed on placebo (Montgomery *et al.* 1988; Rouillon *et al.* 1989; Frank *et al.* 1990), the initiation and maintenance of prophylaxis is highly recommended for those with recent unipolar illness. As noted above, this might achieve a dual benefit of not only preventing the recurrence of expected episodes (with their morbidity and potential for mortality), but may, as well, prevent the sensitization phenomena and cycle acceleration that can occur with these recurrences (and the potential for chronicity or refractoriness) (Fig. 10.5). Episodes not only represent acute biological phenomena but may leave behind longer-lasting vulnerability to recurrence based on changes in gene expression and their downstream effects, which can extend into the well interval, such as persistent cortisol hypersecretion, blunted TSH response to TRH, blunted GH response to clonidine, abnormal reduction in slow wave sleep (Post 1992). In addition, repeated episodes of cortisol hypersecretion interacting with increasing vulnerability due to ageing could damage neurons in the hippocampus (Sapolsky 1990; McEwen 1991), rendering older patients with affective disorder vulnerable to persistent cognitive defects (Rubinow *et al.* 1984). Furthermore, early effective

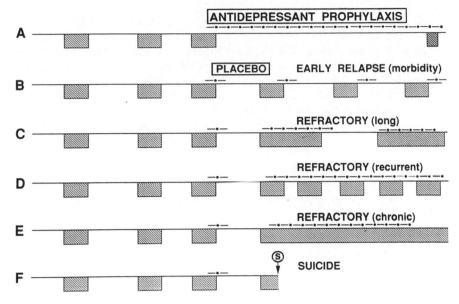

FIG. 10.5. Consequences of failure to prophylax recurrent depression.

prophylaxis might also prevent the transition to treatment-refractoriness, should this occur as a function of repeated drug discontinuations in the unipolar illnesses, as we have seen in a small subgroup of patients with bipolar disorders.

REFERENCES

Angst, J. (1986). The course of affective disorders. *Psychopathology*, **19**, 47–52.

Angst, J. and Weiss, P. (1967). Periodicity of depressive psychoses. In *Proceedings of the fifth international medical foundation; international congress series* (ed. H. Brill, J.O. Cole, and P. Deniker), pp. 702–10. Excerpta Medica, Amsterdam.

Angst, J., Merikangas, K., Scheidegger, P., and Wicki, W. (1990). Recurrent brief depression: a new subtype of affective disorder. *Journal of Affective Disorders*, **19**, 87–98.

Anokhin, K.V. and Rose, S.P.R. (1991). Learning-induced increase of immediate early gene messenger RNA in the chick forebrain. *European Journal of Neuroscience*, **3**, 162–7.

Bauer, M.S. and Whybrow, P.C. (1991). Rapid cycling bipolar disorder: clinical features, treatment, and etiology. In *Advances in neuropsychiatry and psychopharmacology, Vol. 2. Refractory depression* (ed. J.D. Amsterdam), pp. 191–208. Raven Press. New York.

Berridge, M.J. (1989). The Albert Lasker Medical Awards. Inositol trisphosphate, calcium, lithium, and cell signaling. *Journal of the American Medical Association*, **262**, 1834–41.

Bittman, B.J. and Young, R.C. (1991). Mania in an elderly man treated with bupropion. *American Journal of Psychiatry*, **148**, 541.

Bjork, K. (1983). The efficacy of zimeldine in preventing depressive episodes in recurrent major depressive disorders—a double-blind placebo-controlled study. *Acta Psychiatrica Scandinavica* (Suppl. 68), 182–9.

Bratfos, O. and Haug, J.O. (1968). The course of manic-depressive psychosis: follow-up investigation of 215 patients. *Acta Psychiatrica Scandinavica*, **44**, 89–112.

Bunney, W.E., Jr and Garland-Bunney, B.L. (1987). Mechanisms of action of lithium in affective illness: basic and clinical implications. In *Psychopharmacology: the third generation of progress* (ed. H.Y. Meltzer), pp. 553–65. Raven Press, New York.

Bunney, W.E., Jr, Murphy, D.L., Goodwin, F.K., and Borge, G.F. (1970). The switch process from depression to mania: relationship to drugs which alter brain amines. *Lancet*, **1**, 1022–7.

Cavazos, J.E. and Sutula, T.P. (1990). Progressive neuronal loss induced by kindling: a possible mechanism for mossy fiber synaptic reorganization and hippocampal sclerosis. *Brain Research*, **527**, 1–6.

Clark, M., Post, R.M., Weiss, S.R.B., Cain, C.J., and Nakajima, T. (1991). Regional expression of c-fos mRNA in rat brain during the evolution of amygdala kindled seizures. *Molecular Brain Research*, **11**, 55–64.

Coppen, A., Ghose, K., Montgomery, S., Rama Rao, V.A., Bailey, J., and Jorgensen, A. (1978). Continuation therapy with amitriptyline in depression. *British Journal of Psychiatry*, **133**, 28–33.

Crabtree, G.R. (1989). Contingent genetic regulatory events in T lymphocyte activation. *Science*, **243**, 355–61.

Cutler, N.R. and Post, R.M. (1982). Life course of illness in untreated manic-depressive patients. *Comprehensive Psychiatry*, **23**, 101–15.

Daval, J.L., Nakajima, T., Gleiter, C.H., Post, R.M., and Marangos, P.J. (1989). Mouse brain c-fos mRNA distribution following a single electroconvulsive shock. *Journal of Neurochemistry*, **52**, 1954–7.

Doogan, D.P. and Caillard, V. (1992). Sertraline in the prevention of depression. *British Journal of Psychiatry*, **160**, 217–22.

Dragunow, M. and Faull, R.L. (1989). Rolipram induces c-fos protein-like immunoreactivity in ependymal and glial-like cells in adult rat brain. *Brain Research*, **501**, 382–8.

Ellicott, A., Hammen, C., Gitlin, M., Brown, G., and Jamison, K. (1990). Life events and the course of bipolar disorder. *American Journal of Psychiatry*, **147**, 1194–8.

Fabre, L.F. (1991). Buspirone in the treatment of major depression. *Presentation, Fabre Clinic, Houston, TX*.

Faedda, G., Tondo, L., Baldessarini, R.J., Suppes, T., and Tohen, M. (1993). Outcome after rapid vs gradual discontinuation of lithium treatment in bipolar disorders. *Archives of General Psychiatry*, **50**, 448–55.

Fontana, D.J., Post, R.M., and Pert, A. (1991). Conditioned increases in mesolimbic dopamine overflow by stimuli associated with cocaine. *Abstracts, Society for Neuroscience.*

Fontana, D.J., Post, R.M., and Pert, A. (1993). Conditioned increases in meso-limbic dopamine overflow by stimuli associated with cocaine. *Brain Research.* (In press.)

Frank, E., Kupfer, D.J., Perel, J.M., Cornes, C., Jarrett, D.B., Mallinger, A.G. *et al.* (1990). Three-year outcomes for maintenance therapies in recurrent depression. *Archives of General Psychiatry*, **47**, 1093–9.

Gelenberg, A.J., Kane, J.M., Keller, M.B., Lavori, P., Rosenbaum, J.F., Cole, K., and Lavelle, J. (1989). Comparison of standard and low serum levels of lithium for maintenance treatment of bipolar disorder. *New England Journal of Medicine*, **321**, 1489–93.

Georgotus, A., McCue, R.E., and Cooper T.B. (1989) A placebo-controlled comparison of nortriptyline and phenelzine in maintenance therapy of elderly depressed patients. *Archives of General Psychiatry*, **46**, 783–6.

Glen, A.I.M., Johnson, A.L., and Shepherd, M. (1984). Continuation therapy with lithium and amitriptyline in unipolar depressive illness: a randomized, double-blind, controlled trial. *Psychological Medicine*, **14**, 37–50.

Goddard, G.V. and Douglas, R.M. (1975). Does the engram of kindling model the engram of normal long-term memory. *Canadian Journal of Neurological Science*, **2**, 385–95.

Goddard, G.V., McIntyre, D.C., and Leech, C.K. (1969). A permanent change in brain function resulting from daily electrical stimulation. *Experimental Neurology*, **25**, 295–330.

Goodwin, F.K. and Jamison, K.R. (1984). The natural course of manic-depressive illness. In *Neurobiology of mood disorders* (ed. R.M. Post and J.C. Ballenger) pp. 20–37. Williams & Wilkins, Baltimore.

Graybiel, A.M., Moratalla, R., and Robertson, H.A. (1990). Amphetamine and cocaine induce drug-specific activation of the c-fos gene in striosome-matrix compartments and limbic subdivisions of the striatum. *Proceedings of the National Academy of Sciences, USA*, **87**, 6912–16.

Grof, P., Angst, J., and Haines, T. (1974). The clinical course of depression: practical issues. In *Classification and prediction of outcome of depression*, Symposia Medica Hoechst, Vol. 8 (ed. J. Angst), pp. 141–8. F.K. Schattauer, Stuttgart.

Haykal, R.F. and Akiskal, H.S. (1990). Bupropion as a promising approach to rapid cycling bipolar II patients. *Journal of Clinical Psychiatry*, **51**, 450–5.

Himmelhoch, J.M., Thase, M.E., Mallinger, A.G., and Houck, P. (1991). Tranyl-cypromine versus imipramine in anergic bipolar depression. *American Journal of Psychiatry*, **148**, 910–16.

Jakovljevic, M. and Mewett, S. (1991). Comparison between paroxctine, imipramine and placebo in preventing recurrent major depressive episodes. *European Neuropsychopharmacology*, **1**, 440.

Kane, J.M., Quitkin, F.M., Rifkin, A., Ramos-Lorenzi, J.R., Nayak, D.D., and Howard, A. (1982). Lithium carbonate and imipramine in the prophylaxis of unipolar and bipolar II illness. *Archives of General Psychiatry*, **39**, 1065–9.

Kato, N., Higuchi, T., and Friesen, H.G. (1983). Changes of immunoreactive somatostatin and beta-endorphin content in rat brain after amygdala kindling. *Life Sciences*, **32**, 2415–22.

Kraepelin, E. (1921). *Manic-depressive insanity and paranoia* (ed. G.M. Robertson) (trans. R.M. Barclay). E.S. Livingstone, Edinburgh.

Kramlinger, K.G. and Post, R.M. (1988). Ultra-rapid cycling bipolar affective disorder. *Presentation, Annual Meeting, Psychiatric Research Society*, Park City, Utah, March 10–12.

Kubek, M.J., Fuson, K.S., Aydelotte, M.R., and Knoblach, S.M. (1990). Kindling increases neuronal production of thyrotropin-releasing hormone mRNA in seizure foci as determined by *in situ* hybridization histochemistry. *Epilepsia*, **31**, 663.

Kukopulos, A., Reginaldi, D., Laddomada, P., Floris, G., Serra, G., and Tondo, L. (1980). Course of the manic-depressive cycle and changes caused by treatment. *Pharmakopsychiatrie Neuro-psychopharmakologie*, **13**, 156–67.

Kukopulos, A., Caliari, B., Tundo, A., Minnai, G., Floris, G., Reginaldi, D. *et al.* (1983). Rapid cyclers, temperament, and antidepressants. *Comprehensive Psychiatry*, **24**, 249–58.

Kupfer, D.J., Carpenter, L.L., and Frank, E. (1988). Possible role of antidepressants in precipitating mania and hypomania in recurrent depression. *American Journal of Psychiatry*, **145**, 804–8.

Lewis, J. and Winokur, G. (1989). Induction of mania by antidepressants. (Letter.) *American Journal of Psychiatry*, **146**, 126–8.

Lewis, J.L. and Winokur, G. (1982). The induction of mania. A natural history study with controls. *Archives of General Psychiatry*, **39**, 303–6.

Lucibello, F.C., Slater, E.P., Jooss, K.U., Beato, M., and Muller, R. (1990). Mutual transrepression of Fos and the glucocorticoid receptor: involvement of a functional domain in Fos which is absent in Fos B. *EMBO Journal*, **9**, 2827–34.

Lundqvist, G. (1945). Prognosis and course in manic-depressive psychosis. *Acta Psychiatrica et Neurologica Scandinavica*, **35**, 1–96.

McEwen, B.S. (1991). Glucocorticoid actions in the hippocampus: implications for affective disorders. Presentation, *Yale symposium on neurobiology of affective disorders*.

Millbrandt, J. (1991). NGF induces transcription of genes encoding zinc-finger proteins. Abstracts, *American association for the advancement of science annual meeting*.

Mishkin, M. and Appenzeller, T. (1987). The anatomy of memory. *Scientific American*, **256**, 80–9.

Montgomery, S.A. and Dunbar, G.C. (1991). Paroxetine and placebo in the long-term maintenance of depressed patients. *ACNP Abstracts*, p. 117.

Montgomery, S.A., Dufour, H., Brion, S., Gailledreau, J., Laqueille, X., Ferrey, G. *et al.* (1988). The prophylactic efficacy of fluoxetine in unipolar depression. *British Journal of Psychiatry*, **153**, 69–76.

Morgan, J.I. and Curran, T. (1991). Stimulus-transcription coupling in the nervous system: involvement of the inducible proto-oncogenes fos and jun. *Annual Review of Neuroscience*, **14**, 421–51.

Nakajima, T., Daval, J.L., Gleiter, C.H., Deckert, J., Post, R.M., and Marangos, P.J. (1989a). C-fos mRNA expression following electrical-induced seizure and acute nociceptive stress in mouse brain. *Epilepsy Research*, **4**, 156–9.

Nakajima, T., Post, R.M., Weiss, S.R.B., Pert, A., and Ketter, T. (1989*b*). Perspectives on the mechanism of action of electroconvulsive therapy: anticonvulsant, dopaminergic, and c-fos oncogene effects. *Convulsive Therapy*, 5, 274–95.

O'Connell, R.A., Mayo, J.A., Flatow, L., Cuthbertson, B., and O'Brien, B.E. (1991). Outcome of bipolar disorder on long-term treatment with lithium. *British Journal of Psychiatry*, 159, 123–9.

Paskind, H.A. (1930). Manic-depressive psychosis in a private practice. *Archives of Neurology and Psychiatry*, 23, 699–794.

Paykel, E.S. (1979). Causal relationship between clinical depression and life events. In *Stress and mental disorder* (ed. J.E. Barrett, R.M. Rose, and G.L. Klerman), pp. 71–86. Raven Press, New York.

Paykel, E.S. (1982). Life events and early environment. In *Handbook of affective disorders* (ed. E.S. Paykel), pp. 146–61. Churchill Livingstone, New York.

Pazzaglia, P.J. and Post, R.M. (1992). Contingent tolerance and re-response to carbamazepine: a case study in a patient with trigeminal neuralgia and bipolar disorder. *Journal of Neuropsychiatry and Clinical Neuroscience* 4, 76–81.

Pickar, D., Murphy, D.L., Cohen, R.M., Campbell, I.C., and Lipper, S. (1982). Selective and nonselective monoamine oxidase inhibitors: behavioural disturbances during their administration to depressed patients. *Archives of General Psychiatry*, 39, 535–40.

Pinel, J.P.J. (1981). Kindling-induced experimental epilepsy in rats: cortical stimulation. *Experimental Neurology*, 72, 559–69.

Pinel, J.P.J. and Rovner, L.I. (1978*a*). Experimental epileptogenesis: kindling-induced epilepsy in rats. *Experimental Neurology*, 58, 190–202.

Pinel, J.P.J. and Rovner, L.I. (1978*b*). Electrode placement and kindling-induced experimental epilepsy. *Experimental Neurology*, 58, 335–46.

Post, R.M. (1987). Mechanisms of action of carbamazepine and related anticonvulsants in affective illness. In *Psychopharmacology: a generation of progress* (ed. H. Meltzer and W.E. Bunney, Jr), pp. 567–76. Raven Press, New York.

Post, R.M. (1988). Time course of clinical effects of carbamazepine: implications for mechanisms of action. *Journal of Clinical Psychiatry*, 49, 35–46.

Post, R.M. (1990*a*). Prophylaxis of bipolar affective disorders. *International Review of Psychiatry*, 2, 277–320.

Post, R.M. (1990*b*). Alternatives to lithium for bipolar affective illness. In *Review of psychiatry*, Vol. 9 (ed. A. Tasman, S.M. Goldfinger, and C.A. Kaufmann), pp. 170–202. American Psychiatric Press, Washington, DC.

Post, R.M. (1992). The transduction of psychosocial stress into the neurobiology of recurrent affective disorder. *American Journal of Psychiatry*, 149, 999–1010.

Post, R.M. and Weiss, S.R.B. (1989). Non-homologous animal models of affective illness: clinical relevance of sensitization and kindling. In *Animal models of depression* (ed. G. Koob, C. Ehlers, and D.J. Kupfer), pp. 30–54. Birkhauser Boston, Boston.

Post, R.M., Ballenger, J.C., Rey, A.C., and Bunney, W.E., Jr (1981*a*). Slow and rapid onset of manic episodes: implications for underlying biology. *Psychiatry Research* 4, 229–37.

Post, R.M., Lockfeld, A., Squillace, K.M., and Contel, N.R. (1981*b*). Drug-environment interaction: context dependency of cocaine-induced behavioral sensitization. *Life Sciences* **28**, 755–60.

Post, R.M., Rubinow, D.R., and Ballenger, J.C. (1984). Conditioning, sensitization, and kindling: implications for the course of affective illness. In *Neurobiology of mood disorders* (ed. R.M. Post and J.C. Ballenger), pp. 432–66. Williams & Wilkins, Baltimore.

Post, R.M., Rubinow, D.R., and Ballenger, J.C. (1986). Conditioning and sensitization in the longitudinal course of affective illness. *British Journal of Psychiatry*, **149**, 191–201.

Post, R.M., Weiss, S.R.B., and Pert, A. (1987). The role of context in conditioning and behavioral sensitization to cocaine. *Psychopharmacology Bulletin*, **23**, 425–9.

Post, R.M., Rubinow, D.R., Uhde, T.W., Roy-Byrne, P.P., Linnoila, M., Rosoff, A. *et al.* (1989). Dysphoric mania: clinical and biological correlates. *Archives of General Psychiatry*, **46**, 353–8.

Post, R.M., Kramlinger, K.G., Altshuler, L.L., Ketter, T., and Denicoff, K. (1990*a*). Treatment of rapid cycling bipolar illness. *Psychopharmacology Bulletin*, **26**, 37–47.

Post, R.M., Leverich, G., Rosoff, A.S., and Altshuler, L.L. (1990*b*). Carbamazepine prophylaxis in refractory affective disorders. A focus on long-term followup. *Journal of Clinical Psychopharmacology*, **10**, 318–27.

Post, R.M., Weiss, S.R.B., Clark, M., Nakajima, T., and Pert, A. (1990*c*). Amygdala versus local anesthetic kindling: differential anatomy, pharmacology, and clinical implications. In *Kindling IV* (ed. J. Wada), pp. 357–69. Plenum Press, New York.

Post, R.M., Weiss, S.R.B., Nakajima, T., Clark, M. and Pert, A. (1990*d*). Mechanism-based approaches to anticonvulsant therapy. In *Current and future trends in anticonvulsant, anxiety and stroke therapy* (ed. B.S. Meldrum and M. Williams), pp. 45–90. Wiley-Liss, New York.

Post, R.M., Weiss, S.R.B., and Pert, A. (1991*a*). Animal models of mania. In *The mesolimbic dopamine system: from motivation to action* (ed. P. Willner and J. Scheel-Kruger), pp. 443–72. Wiley, Chicester.

Post, R.M., Weiss, S.R.B., Ketter, T., Pazzaglia, P.J., and Denicoff, K. (1991*b*). Impact of affective illness on gene expression: rationale for long-term prophylaxis. *European Neuropsychopharmacology*, **1**, 214–16.

Post, R.M., Weiss, S.R.B., Uhde, T.W., Clark, M., and Rosen, J.B. (1992*a*). Preclinical neuroscience advances pertinent to panic disorder: implications of cocaine kindling, induction of the proto-oncogene c-fos, and contingent tolerance. In *Biology of anxiety disorders: recent developments* (ed. R. Hoehn-Saric). American Psychiatric Association Press, Washington, DC.

Post, R.M., Weiss, S.R.B., and Chuang, D.-M. (1992*b*). Mechanisms of action of anticonvulsants in affective disorders: comparisons with lithium. *Journal of Clinical Psychopharmacology*, **12**, 23S–35S.

Post, R.M., Leverich, G., Altshuler, L., and Mikalauskas, K. (1992*c*). Lithium discontinuation-induced refractoriness: preliminary observations. *American Journal of Psychiatry*, **149**, 1727–9.

Potter, W.Z., Murphy, D.L., Wehr, T.A., Linnoila, M., and Goodwin, F.K.

(1982). Clorgyline. A new treatment for patients with refractory rapid-cycling disorder. *Archives of General Psychiatry*, **39**, 505–10.

Prien, R.F., Klett, C.J., and Caffey, E.M., Jr (1973). Lithium carbonate and imipramine in prevention of affective episodes. A comparison in recurrent affective illness. *Archives of General Psychiatry*, **29**, 420–5.

Prien, R.F., Kupfer, D.J., Mansky, P.A., Small, J.G., Tuason, V.B., Voss, C.B. *et al.* (1984). Drug therapy in the prevention of the recurrences in unipolar and bipolar affective disorders. *Archives of General Psychiatry*, **41**, 1096–104.

Racine, R. (1978). Kindling: the first decade. *Neurosurgery*, **3**, 234–52.

Robertson, G.S., Vincent, S.R., and Fibiger, H.C. (1990). Striatonigral projection neurons contain D1 dopamine receptor-activated c-fos. *Brain Research*, **523**, 288–90.

Robertson, H.A., Peterson, M.R., Murphy, K., and Robertson, G.S. (1989). D1-Dopamine receptor agonists selectively activate striatal c-fos independent of rotational behaviour. *Brain Research*, **503**, 346–9.

Robinson, D.S., Alms, D.R., Shrotriya, R.C., Messina, M., and Wickramaratne, P. (1989). Serotonergic anxiolytics and treatment of depression. *Psychopathology*, **22** 27–36.

Robinson, D.S., Lerfald, S.C., Bennett, B., Laux, D., Devereaux, E., Kayser, A. *et al.* (1991). Continuation and maintenance treatment of major depression with the monoamine oxidase inhibitor phenelzine: a double-blind placebo-controlled discontinuation study. *Psychopharmacology Bulletin*, **27**, 31–9.

Rose, S.P.R. (1991). How chicks make memories: the cellular cascade from c-fos to dendritic remodelling. *Trends in Neuroscience*, **14**, 390–7.

Rosen, J.B., Abramowitz, J., and Post, R.P. (1993). Co-localization of TRH mRNA and fos-like immunoreactivity in limbic structures following amygdala kindling. *Molecular and Cellular Neurosciences*, **4**, 335–42.

Rouillon, F. (1991). Adverse drug reaction in long term antidepressant treatment. *European Neuropsychopharmacology*, **1**, 216–18.

Rouillon, F., Phillips, R., Serrurier, D., Ansart, E., and Gerard, M.J. (1989). (Recurrence of unipolar depression and efficacy of maprotiline). *Encephale*, **15**, 527–34.

Roy-Byrne, P.P., Post, R.M., Uhde, T.W., Porcu, T., and Davis, D.D. (1985). The longitudinal course of recurrent affective illness: life chart data from research patients at NIMH. *Acta Psychiatrica Scandinavica*, (Suppl.) **71**, 5–34.

Rubinow, D.R., Post, R.M., Savard, R., Gold, P.W. (1984). Cortisol hypersecretion and cognitive impairment in depression. *Archives of General Psychiatry*, **41**, 279–83.

Sapolsky, R.M. (1990). Adrenocortical function, social rank, and personality among wild baboons. *Biological Psychiatry*, **28**, 862–78.

Schou, M. (1979). Lithium as a prophylactic agent in unipolar affective illness. *Archives of General Psychiatry*, **36**, 849–51.

Shopsin, B. (1983). Bupropion's prophylactic efficacy in bipolar affective illness. *Journal of Clinical Psychiatry*, **44**, 163–9.

Shuman, J.D., Vinson, C.R., and McKnight, S.L. (1990). Evidence of changes in protease sensitivity and subunit exchange rate on DNA binding by C/EBP. *Science*, **249**, 771–3.

Smith, M., Weiss, S.R.B., Abedin, T., Post, R.M., and Gold, P. (1991). Effects of amygdala-kindling and electroconvulsive seizures on the expression of cortico-tropin releasing hormone (CRH) mRNA in the rat brain. *Molecular and Cellular Neurosciences*, **2**, 103–16.

Sonnenberg, J.L., Rauscher, F.J., III, Morgan, J.I., and Curran, T. (1989). Regulation of proenkephalin by Fos and Jun. *Science*, **246**, 1622–5.

Souza, F.G., Mander, A.J., and Goodwin, G.M. (1990). The efficacy of lithium in prophylaxis of unipolar depression: evidence from its discontinuation. *British Journal of Psychiatry*, **157**, 718–22.

Squillace, K., Post, R.M., Savard, R., and Erwin, M. (1984). Life charting of the longitudinal course of recurrent affective illness. In *Neurobiology of mood disorders* (ed. R.M. Post and J.C. Ballenger), pp. 38–59. Williams & Wilkins, Baltimore.

Squire, L.R. and Zola-Morgan, S. (1991). The medial temporal lobe memory system. *Science*, **253**, 1380–6.

Strober, M.T., Morrell, W., Lampert, C., and Burroughs, J. (1990). Relapse following discontinuation of lithium maintenance therapy in adolescents with bipolar I illness: a naturalistic study. *American Journal of Psychiatry*, **147**, 457–61.

Suppes, T., Tohen, M., and Faedda, G. (1991). Risk of discontinuation of lithium in bipolar disorder. *Abstracts, American Psychiatric Association*, **61**, Abs. No. NR52.

Sutula, T.P. (1990). Experimental models of temporal lobe epilepsy: new insights from the study of kindling and synaptic reorganization. *Epilepsia*, **31**, S45–S54.

Swift, J.M. (1907). The prognosis of recurrent insanity of the manic-depressive type. *American Journal of Insanity*, **64**, 311–26.

Taschev, T. (1974). The course and prognosis of depression on the basis of 652 patients decreased. In *Symposia Medica Hoest: classification and prediction of outcome of depression* (ed. F.K. Schattauer), pp. 156–72. Schattauer, New York.

Vinson, C.R., Sigler, P.B., and McKnight, S.L. (1989). Scissors-grip model for DNA recognition by a family of leucine zipper proteins. *Science*, **246**, 911–16.

Wada, J.A., Sato, M., and Corcoran, M.E. (1974). Persistent seizure susceptibility and recurrent spontaneous seizures in kindled cats. *Epilepsia*, **15**, 465–78.

Wehr, T.A. and Goodwin, F.K. (1987a). Do antidepressants cause mania? *Psychopharmacology Bulletin*, **23**, 61–5.

Wehr, T.A. and Goodwin, F.K. (1987b). Can antidepressants cause mania and worsen the course of affective illness? *American Journal of Psychiatry*, **144**, 1403–11.

Wehr, T.A., Sack, D.A., Rosenthal, N.E., and Cowdry, R.W. (1988). Rapid cycling affective disorder: contributing factors and treatment responses in 51 patients. *American Journal of Psychiatry*, **145**, 179–84.

Weiss, S.R.B. and Post, R.M. (1991). Contingent tolerance to carbamazepine: a peripheral-type benzodiazepine mechanism. *European Journal of Pharmacology*, **193**, 159–63.

Weiss, S.R.B., Post, R.M., Pert, A., Woodward, R., and Murman, D. (1989). Context-dependent cocaine sensitization: differential effect of haloperidol on development versus expression. *Pharmacology, Biochemistry and Behavior*, **34**, 655–61.

Yang-Yen, H.F., Chambard, J.C., Sun, Y.L., Smeal, T., Schmidt, T.J., Drouin, J. *et al.* (1990). Transcriptional interference between c-Jun and the glucocorticoid receptor: mutual inhibition of DNA binding due to direct protein-protein interaction. *Cell*, **62**, 1205–15.

Young, S.T., Porrino, L.J., and Iadarola, M.J. (1989). Induction of c-fos by direct and indirect dopamine agonists. *Abstracts, Society for Neuroscience*, **15**. Abs. No. 432.5–1091.

Young, S.T., Porrino, L.J., and Iadarola, M.J. (1991). Cocaine induces striatal c-fos immunoreactive proteins via dopaminergic D1 receptors. *Proceedings of the National Academy of Sciences*, **88**, 1291–5.

Zis, A.P. and Goodwin, F.K. (1979). Major affective disorder as a recurrent illness. *Archives of General Psychiatry*, **36**, 835–9.

Zis, A.P., Grof, P., and Webster, M. (1980). Prediction of relapse in recurrent affective disorder. *Psychopharmacology Bulletin*, **16**, 47–9.

Zubicta, J.K. and Demitrack, M.A. (1991). Possible buproprin precipitation of mania and a mixed affective state. (Letter.) *Journal of Clinical Psychopharmacology*, **11**, 327.

11

Testing a proposed model on central serotonergic function and impulsivity

MARKKU LINNOILA and MATTI VIRKKUNEN

In a series of studies, we have previously observed that Finnish impulsive violent offenders and fire setters have relatively low CSF 5-hydroxy-indoleacetic acid (5-HIAA) concentrations (Linnoila *et al.* 1983; Virkkunen *et al.* 1987). They also experience mild hypoglycaemic episodes during oral glucose tolerance tests and sleep irregularly while on the forensic psychiatry ward (Roy *et al.* 1986). Based on these observations we have proposed a model in which deficient central serotonin turnover in impulsive violent offenders is conducive of disturbances of diurnal activity rhythm and glucose metabolism (Fig. 11.1) (Linnoila *et al.* 1986). The neuroanatomical

PROPOSED PATHOGENESIS OF IMPULSIVITY

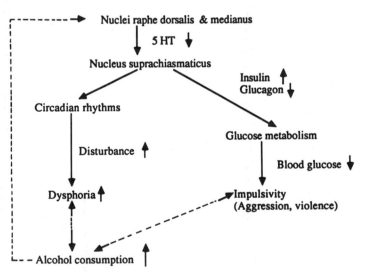

FIG. 11.1. Model to explain relationships between central serotonin turnover, diurnal rhythms, glucose metabolism, mood, impulse control, and alcohol consumption. For more detailed discussion see text.

substrate postulated to have an important role in the model is the supra-chiasmatic nucleus, which receives a serotonergic input from the dorsal and median raphe nuclei (Palkovits *et al.* 1977) and functions as a circadian pacemaker (Moore *et al.* 1972) and also as a regulator of glucose metabolism (Yamamoto *et al.* 1984).

In follow up and family history studies on offenders, we have found that a low CSF 5-HIAA concentration and propensity to mild hypoglycaemias are predictive of recidivist violent criminality after release from prison (Virkkunen *et al.* 1989*a*). Suicide attempts and successful suicides are primarily associated with low 5-HIAA and 3-methoxy-4-hydroxyphenyl-glycol (MHPG) concentrations in the CSF (Virkkunen *et al.* 1989*b*). Moreover, sons of alcoholic fathers, who have been convicted of violent crimes, have the lowest CSF 5-HIAA concentrations (Linnoila *et al.* 1989). Here we report results which directly test several aspects of the proposed model. Furthermore, for the first time, we investigate age- and sex-matched Finnish healthy volunteers as inpatients in the same psychiatry department as the violent offenders. The *a priori* hypotheses explored are (1) low CSF 5-HIAA concentration is associated with impulsivity of the index crime, a history of suicide attempts, a disturbance of diurnal activity rhythms, and abnormalities in glucose metabolism and (2) CSF-free testosterone concentration is correlated with aggressiveness rather than impulsiveness.

MATERIALS AND METHODS

Subjects and procedures

Healthy volunteers were recruited by advertisements. They were asked to participate in a psychobiological study elucidating biochemical, behavioural, and genetic correlates of violent behaviour under the influence of alcohol. The advertisement further defined that the volunteers had to be free of current or past drinking problems and mental disorders. After recruitment, the volunteers gave a venous blood sample to establish that they had normal blood count, liver enzymes, and serum creatinine concentration. They were asked to follow a low monoamine and caffeine-free diet, not to use any alcohol or medicines for a week prior to admission, and to stay as inpatients on the research ward for 3 days and 3 nights.

On the first full day after a night on the ward, they underwent a lumbar puncture between 8 and 9 a.m. On the next two days they received double-blind, random order, oral glucose or aspartame tests starting at 8 a.m.

Violent offenders and impulsive fire setters were ordered to undergo forensic psychiatric examinations by their trial judges. They spent an average of 1 month on the low monoamine diet, drug free in the research ward. A physician not involved in the study asked the offenders to volunteer, and explained the procedures involved. All subjects gave a written informed consent. The study protocol was approved by the Department of Psychiatry and Helsinki University Central Hospital Institutional Review Boards in Finland. The Helsinki University Central Hospital Review Board had an *ad hoc* prisoner representative to evaluate the protocol. Gamma glutamyl transpeptidase (GGT) was quantified in blood prior to the onset of the study to ascertain maintenance of abstinence from alcohol. The offenders wore the physical activity monitors on their left wrists continuously for 10 days and nights. They underwent double-blind, random order oral glucose and aspartame tests on 2 consecutive days.

Psychiatric diagnoses and family history

All subjects were administered an SADS-L (Spitzer and Endicott 1978) and a clinical interview (by M.V.) to derive RDC and DSM-III-R (APA 1978) diagnoses, and to elicit a family history of psychiatric and substance-abuse disorders, alcoholism, and attempted or successful suicide. To maintain continuity, intermittent explosive disorder was diagnosed according to DSM-III criteria which unlike DSM-III-R permit the diagnosis when the behaviour is exhibited under the influence of alcohol.

Biochemical variables

Monoamine metabolites

The CSF samples were obtained at 8.00 a.m. after one night of bed rest with only water permitted after 8.00 p.m. The samples were collected into a large polypropylene tube on wet ice. After the first 12 ml had been drawn, the tube was capped, inverted, and the CSF was aliquotted into 1 ml tubes on dry ice. Homovanillic acid (HVA), 5-HIAA, and MHPG concentrations were quantified with a liquid chromatographic procedure using electrochemical detection (Scheinin *et al.* 1983).

CSF-free testosterone

CSF-free testosterone concentration was quantified with radioimmunoassay (Rahe *et al.* 1990).

Oral glucose and aspartame tolerance tests

After a 12-h overnight fast, at 8.00 a.m. the subjects consumed 1 g/kg of body weight (4 ml/kg) of glucose solution or an identical volume of an aspartame solution of indistinguishable sweetness (Leiras, Turku, Finland). Fifteen millilitre blood samples were drawn from an antecubital vein into an aprotinin containing test-tube (12.5 mIU/ml, Antagosan®

Behringwerke, Marburg, Germany) prior to and 15, 30, 60, 90, 120, 180, 240, and 300 min after the administration of the liquid. For the first 2 h of the test, the subjects rested in bed. Thereafter, they were allowed to move on the ward, but resting was encouraged. Blood glucose concentration was measured using a glucose dehydrogenase method (Banauch *et al.* 1975). Insulin was quantified in antibody coated test-tubes (Coat-A-Count®; Diagnostic Products Corporation, Los Angeles, USA). Between-assay variation at 30.2 uU/ml for insulin was 4.6 per cent.

Physical activity monitoring

The activity monitors are small watch-size devices which have a movement sensor, as well as clock and memory functions which permit continuous recording of activity for a period of 10 days (Wehr *et al.* 1982). The data were decoded and stored on an Apple Macintosh computer.

Data analyses

All analyses were computed using the BMDP statistical package (BMDP Statistical Software Manual 1990). Parametric and nonparametric analyses of variance, correlation coefficients, and *post hoc* tests for comparisons of individual means were computed when appropriate. Bonferroni correction was used for multiple comparisons. Two-tailed probabilities were applied except when clearly dimensional *a priori* hypotheses were tested as stated in the introduction. Preliminary analyses revealed no differences between the various groups in mean age, weight, height, or baseline glucose, and insulin concentrations. Also age, weight, or height did not contribute significantly to the variation of any of the psychobiological variables in the present sample. Therefore, these variables were not statistically adjusted in any of the analyses reported below. All results are expressed as means and standard deviations.

To examine directly biochemical concomitants of impulsiveness a stepwise linear discriminant analysis was computed on the CSF biochemical variables for the impulsive–nonimpulsive offender grouping. A similar analysis was computed for the offender–healthy volunteer grouping after excluding the fire setters to investigate biochemical concomitants of aggressiveness. The prior probabilities used were 0.5 and 0.5. The *F* in/out value was set at 4.

RESULTS

Demographic characteristics and psychiatric diagnoses

Ages, DSM-III-R diagnoses, intelligence quotients, characteristics of the index crime, and GGT levels are indicated in Table 11.1. The healthy

TABLE 11.1. *Demographic characteristics, intelligence quotients, psychiatric diagnoses, and gamma glutamyl transpeptidase levels (the upper limit of normal for the laboratory is 40) in offenders and healthy volunteers. Healthy volunteers positive denotes volunteers with first-degree relatives with alcohol dependence or depression*

	Impulsive (N = 43)	Nonimpulsive (N = 15)	Healthy volunteers (N = 21)	Healthy volunteers positive (N = 6)
Height (cm)	179.4 ± 4.8	179.3 ± 6.0	179.9 ± 6.5	179.0 ± 1.8
Weight (kg)	76.3 ± 10.3	78.1 ± 12.3	81.8 ± 13.8	72.5 ± 10.6
Age (years)	30.4 ± 9.6	30.2 ± 7.7	27.9 ± 9.1	29.7 ± 8.9
IQ (WAIS, full scale)	102.1 ± 14.2	98.6 ± 20.1	–	–
Antisocial personality	23	0	–	–
Conduct disorder	1	0	–	–
Explosive personality	20	0	–	–
Borderline personality	1	6	–	–
Dysthymic disorder	15	6	–	–
Major depressive disorder	3	4	–	–
Passive–aggressive personality	1	6	–	–
Schizoid personality	1	1	–	–
Narcisstic personality	2	0	–	–
Dependent personality	1	4	–	–
Gamma GTT	22.7 ± 13.3	27.7 ± 5.7	–	–

volunteers were divided into two groups according to their family histories. Twenty-one of them had a family history free of axis I or axis II mental disorders. They are used in the statistical comparisons. Six had first-degree blood relatives with either alcoholism or major depressive disorders. Their data are given separately in the figures and the table, although they are not used in the statistical comparisons due to their low number in the sample. Oversampling of offenders for the day–night activity monitoring, technical difficulties with lumbar punctures and certain assays led to unequal sample sizes in various comparisons. The sample sizes are, therefore, included in all figures and the table.

The offenders were divided into impulsive and nonimpulsive groups based on the characteristics of the index crime as in our previous studies. (A crime was called impulsive when the victim was previously unknown to the offender, when no provocation or only verbal altercation preceded the attack, no premeditation could be documented, and no economic motivation such as robbery or burglary was evident. Impulsive fire setting excluded setting fires for insurance fraud.) (Linnoila *et al.* 1983; Virkkunen *et al.* 1987.) There were ten fire setters in the impulsive offender group. Three of the nonimpulsive offenders had attempted to kill or killed someone by setting their house on fire.

Biochemical variables

CSF monoamine metabolite concentrations

Mean CSF 5-HIAA was significantly lower in the impulsive than the nonimpulsive offenders. The nonimpulsive offenders had significantly higher mean CSF 5-HIAA than the healthy volunteers.

Of the violent offenders, 25 had made a suicide attempt. Their mean CSF 5-HIAA concentration was lower than the nonattempting offenders; CSF 5-HIAA (58.8 ± 25.2 versus 68.5 ± 24.7 nmol/l; $p<0.05$, one-tailed probability).

In the discriminant analysis on the impulsive–nonimpulsive grouping of the offenders the only variable selected with an overall correct jackknife classification rate of 0.78 was CSF 5-HIAA concentration.

Healthy volunteers with a family history positive for first-degree relatives with alcoholism or major depression had CSF 5-HIAA and HVA concentrations similar to that of the impulsive offenders (Figs 11.2 and 11.3).

Mean CSF HVA concentrations were significantly lower in the impulsive than in the nonimpulsive offenders (Fig. 11.3). Mean CSF MHPG concentrations did not differ between any of the groups in the *post hoc* comparisons, even though the ANOVA showed a significant overall difference at the $p<0.05$ level. The impulsive offenders had the lowest and the

CSF 5HIAA

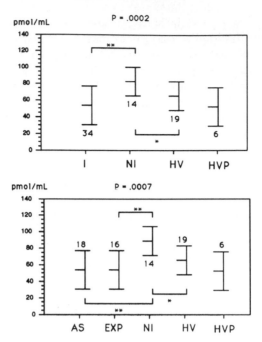

FIG. 11.2. Mean CSF 5-HIAA concentrations. For abbreviations see text. ** = $p < 0.01$; * = $p < 0.05$; number = $p < 0.05$ one-tailed.

healthy volunteers the highest mean CSF MHPG concentrations. The family history positive and negative healthy volunteers had similar mean CSF MHPG concentrations.

CSF-free testosterone concentrations

Mean CSF-free-testosterone concentrations were higher among offenders than healthy volunteers (Fig. 11.4).

In the discriminant analysis on the violent offenders–healthy volunteers grouping the first variable entered was CSF-free testosterone concentration followed by CSF 5-HIAA and MHPG concentrations. The overall correct classification by CFS-free testosterone concentration was 0.60. This was improved to 0.64 by adding the effects of the CSF 5-HIAA and MHPG concentrations.

Oral glucose and aspartame tolerance tests

Impulsive violent offenders had significantly lower mean blood glucose nadir during the glucose tolerance test than healthy volunteers (Fig. 11.5).

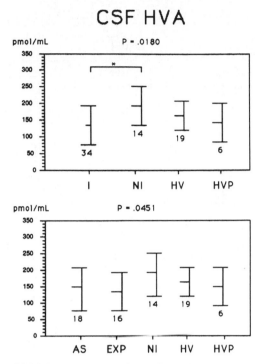

FIG. 11.3. Mean CSF HVA concentrations. For abbreviations see text. For levels of statistical significance see legend for Fig. 11.2.

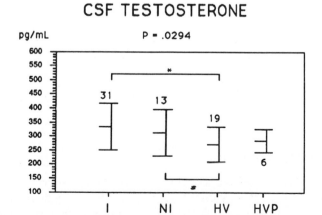

FIG. 11.4. Mean CSF testosterone concentrations. For abbreviations see text. For levels of statistical significance see legend for Fig. 11.2.

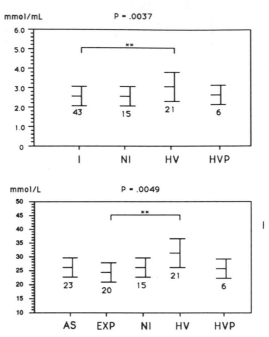

FIG. 11.5. Blood glucose nadir during the first 5 h following an oral (1 g/kg of body weight) glucose challenge. For abbreviations see text. For levels of statistical significance see legend for Fig. 11.2.

Plasma insulin concentrations did not differ significantly between the groups at any time point during the oral glucose tolerance test (data on file). There were no significant differences in any of the biochemical variables at any time points between the groups in the aspartame tolerance test.

Physical activity monitoring

Impulsive offenders with antisocial personality disorder had significantly higher mean total 10 day–night activity counts than healthy volunteers (Fig. 11.6). Impulsive offenders with intermittent explosive disorder had indistinguishable day and night activity counts in a striking difference from the other groups (Fig. 11.6).

PAM PROFILE

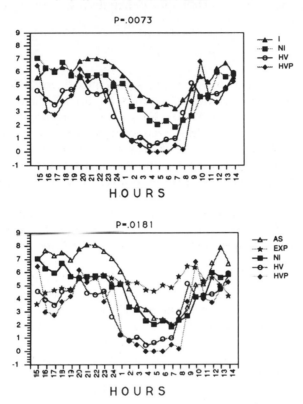

HOURS

FIG. 11.6. Mean hourly physical activity counts. For abbreviations see text.

DISCUSSION

The model proposed to explain poor sleep and hypoglycaemic tendency in impulsive violent offenders may have some validity. The results demonstrate that impulsive violent offenders with the lowest CSF 5-HIAA concentrations have a profound diurnal activity rhythm disturbance. This is particularly true for offenders with the diagnosis of intermittent explosive disorder. In rodents, intact serotonergic input to the suprachiasmatic nucleus facilitates entrainment of circadian rhythms by light (Morin and Blanchard 1991). If the same principle holds for humans, the disturbed diurnal activity rhythm observed in the majority of the impulsive offenders may indeed be secondary to deficient central serotonergic neurotransmission. Compared to the healthy volunteers the offenders with antisocial personality disorder showed increased mean total 24-h activity counts throughout the monitoring period. This finding is commensurate with the history of hyperactivity and attention deficit disorder in many of them (Roy *et al.* 1986).

CSF 5-HIAA

In animal and human studies, indexes of reduced central serotonin turnover have been associated with inability to tolerate delay and impaired impulse control (Soubrie 1986). Somewhat surprisingly a post-mortem human study found that 5-HIAA concentration in the prefrontal cortex correlated positively with CSF 5-HIAA concentration (Stanley *et al.* 1985). This apparent anatomic–neurochemical relationship is intriguing, because of the postulated role of the prefrontal cortex in impulse control (Miller 1992).

The present results replicate our earlier observation (Linnoila *et al.* 1983) of a low mean CSF 5-HIAA concentration in offenders with a positive history of suicide attempts compared to offenders without suicide attempts.

In the present study, even healthy volunteers free of family history of alcoholism or major depression in their first-degree relatives, had relatively low CSF 5-HIAA concentrations. This may be due to not sampling volunteers during late spring and summer when CSF 5-HIAA in volunteers reaches its seasonal peak (Brewerton *et al.* 1988). Additional subjects sampled during all four seasons are needed to permit a truly adequate comparison between volunteers and violent offenders. Our previous studies in alcoholics show a lack of seasonal variation in CSF 5-HIAA and HVA concentrations in this patient group (Roy *et al.* 1991). In accordance with this observation, we again did not find any seasonal variation in CSF monoamine metabolite concentrations in the alcoholic offenders.

Healthy volunteers with a positive family history of alcohol dependence or major depressive disorders in their first-degree relatives had CSF 5-HIAA concentrations similar to the impulsive violent offenders. This finding, although tempered by the low number of subjects, highlights the fact that variables other than a low CSF 5-HIAA are necessary for many of the adverse outcomes associated with this biochemical variable. One such variable may be alcohol and other substance abuse exhibited by all impulsive offenders in the present study. Our previous findings suggest, however, that even in healthy volunteers relatively low CSF 5-HIAA is associated with problems of impulse control, i.e. verbal acting-out hostility (Roy *et al.* 1988).

CSF-free testosterone

Mean CSF-free testosterone concentration was higher in both impulsive and nonimpulsive offenders than in healthy volunteers. It was especially high in the offenders with antisocial personality disorder. The results of the discriminant analyses support the notion that central serotonin turnover is primarily associated with impulse control, and central availability of test-osterone more with outward-directed aggressiveness (Ehrenkranz *et al.* 1976; Dabbs *et al.* 1987).

CSF HVA and MHPG

Differences in the mean CSF HVA concentrations between the groups had a similar profile but were relatively smaller than differences in mean CSF 5-HIAA concentrations. As expected, CSF 5-HIAA and HVA were highly correlated with each other. As we have proposed previously (Agren *et al.* 1986), and as supported by the results of the discriminant analysis, the differences between the groups in mean CSF HVA concentrations are probably secondary to the differences in mean CSF 5-HIAA concentrations.

In general, we have been unable to consistently relate static measures of noradrenergic functions to trait measures of impulsivity and aggressiveness (Roy *et al.* 1989) even though we previously found an association between history positive for attempted suicide and low CSF MHPG concentration in Finnish offenders (Virkkunen *et al.* 1989*b*).

Glucose metabolism

In accordance with Virkkunen's previous findings (Virkkunen *et al.* 1986), mean blood glucose nadir was significantly lower in the impulsive violent offenders than in the healthy volunteers. There was, however, no difference between the impulsive and nonimpulsive offenders in this variable. Thus, the low blood glucose nadir is not correlated to CSF 5-HIAA concentration across the different groups. The difference between the offenders and healthy volunteers could be either secondary to alcohol dependence diagnosed in all but two of the offenders, or due to the different periods of time the two groups were on the low monoamine diet prior to the glucose challenge. No differences between the groups were found in insulin concentrations at any point during the oral glucose tolerance test. To investigate appropriately the glucose metabolism-related aspects of the proposed model on inter-relationships between serotonin turnover, disturbed diurnal rhythm regulation, and glucose metabolism (Fig. 11.1), requires studies utilizing the technique of euglycaemic insulin clamp.

Psychiatric diagnoses

The main difference between our earlier studies and the present findings is that all except two of the nonimpulsive offenders fulfilled criteria for

DSM-III-R alcohol dependence rather than alcohol abuse. This is because the DSM-III-R criteria for alcohol dependence include a number of subjects who previously according to DSM-III were classified as abusers (Cotter *et al.* 1991). The main weakness of the present design is that the diagnoses were made by a forensic research psychiatrist (M.V.) who was familiar with the crimes and the backgrounds of the offenders.

In conclusion, low CSF 5-HIAA concentration seems to be associated with impaired impulse control and day–night activity rhythm dysregulation particularly in offenders with intermittent explosive disorder. Inter-relationships between serotonin and glucose metabolism dysregulation need further study to elucidate the basic physiology involved. Equally low CSF 5-HIAA concentrations can be found in impulsive, alcohol-dependent, violent offenders and in some healthy volunteers with a family history positive for alcoholism and major depressive disorders in first-degree relatives. CSF-free testosterone concentration is associated with outward-directed aggressiveness.

ACKNOWLEDGEMENTS

The authors are grateful to Russell Poland, for the testosterone analyses and Robert Rawlings, for the statistical analyses.

REFERENCES

Agren, H., Mefford, I.N., Rudorfer, M.V., Linnoila, M., and Potter, W.Z. (1986). Interacting neurotransmitter systems: a non-experimental approach to the 5HIAA–HVA correlation in human CSF. *Psychiatry Research*, **20**, 175–93.

APA (American Psychiatry Association) (1978). *Diagnostic and statistical manual* (3rd edn, revised). Washington, DC.

Banauch, D., Brummer, W., Ebeling, W., Metz, H., Rindfrey, H., Lang, H., Leybold, K., and Rick, W. (1975). Eine Glucose-Dehydrogenase fur die Glucose-Bestimmung in Korperflussigkeiten. *Zeitschrist Klinische Chemie Klinische Biochemie*, **13**, 101–7.

BMDP Statistical Software Manual (1990). University of California Press, Los Angeles, CA.

Brewerton, R.D., Berrettini, W.H., Nurnberger, J.I., and Linnoila, M. (1988). Analysis of seasonal fluctuations of CSF monoamine metabolites and neuro-peptides in normal controls. Findings with 5-HIAA and HVA. *Psychiatry Research*, **23**, 257–65.

Cotter, L.B., Helzer, J.E., Mager, D., Spitznagel, E.M., and Compton, W.M. (1991). Agreement between DSM-III and -IIIR substance use disorders. *Drug Alcohol and Dependence*, **29**, 17–25.

Dabbs, J.M., Jr, Frady, R.L., Carr, T.S., and Besch, N.F. (1987). Saliva testosterone and criminal violence in young adult prison inmates. *Psychosomatic Medicine*, **49**, 174–82.

Ehrenkranz, J., Bliss, E., and Sheard, M.H. (1976). Plasma testosterone. Correlation with aggressive behavior and social dominance in man. *Psychosomatic Medicine*, **36**, 469–75.

Linnoila, M., Virkkunen, M., Scheinin, M., Nuutila, A., Rimon, R., and Goodwin, F.K. (1983). Low cerebrospinal fluid 5-hydroxyindoleacetic acid concentration differentiates impulsive from nonimpulsive violent behavior. *Life Science*, **33**, 2609–14.

Linnoila, M., Virkkunen, M., and Roy, A. (1986). Biochemical aspects of aggression in man. In *Clinical neuropharmacology, suppl.1*, (ed. W. E. Bunney, Jr, E. Costa, and S. G. Potkin), pp. 377–9. Raven Press, New York.

Linnoila, M., DeJong, J., and Virkkunen, M. (1989). Family history of alcoholism in violent offenders and impulsive fire setters. *Archives of General Psychiatry*, **46**, 679–81.

Miller, L. A. (1992). Impulsivity, risk-taking, and the ability to synthesize fragmented information after frontal lobectomy. *Neuropsychologia*, **30**, 69–79.

Moore, R.Y. and Eichler, V.B. (1972). Loss of a circadian adrenal corticosterone rhythm following suprachiasmatic lesions in the rat. *Brain Research*, **42**, 201–6.

Morin, L.P. and Blanchard, J. (1991). Depletion of brain serotonin by 5,7-DHT modifies hamster circadian rhythm response to light. *Brain Research*, **566**, 173–85.

Palkovits, M., Saavedra, J.M., Jacobovits, D.M., Kizer, J.S., Zaborsky, L., and Brownstein, M.J. (1977). Serotonergic innervation of the forebrain: effects of lesions on serotonin and tryptophan hydroxylase levels. *Brain Research*, **130**, 121–34.

Rahe, R.H., Karson, S., Howard, N.S., Jr, Rubin, R.T., and Poland, R.E. (1990). Psychological and physiological assessments on American hostages freed from captivity in Iran. *Psychosomatic Medicine*, **52**, 11–16.

Roy, A., Virkkunen, M., Guthrie, S., and Linnoila, M. (1986). Indices of serotonin and glucose metabolism in violent offenders, arsonists and alcoholics. In *Psychobiology of suicidal behavior*, (ed. J.J. Mann and M. Stanley), pp. 202–20. Academy of Science, New York.

Roy, A., Adinoff, B., and Linnoila, M. (1988). Acting out hostility in normal volunteers: negative correlation with CSF-5-HIAA levels. *Psychiatry Research*, **24**, 187–94.

Roy, A., Pickar, D., DeJong, J., Karoum, F., and Linnoila, M. (1989). Suicidal behavior in depression: relationship to noradrenergic function. *Biological Psychiatry*, **25**, 341–50.

Roy, A., Adinoff, B., and Linnoila, M. (1991). Cerebrospinal fluid variables among alcoholics lack seasonal variation. *Acta Psychiatrica Scandinavica*, **24**, 187–94.

Scheinin, M., Chang, W-H., Kirk, K., and Linnoila, M. (1983). Simultaneous determination of 3-methoxy-4-hydroxyphenylglycol, 5-hydroxyindoleacetic acid, and homovanillic acid in cerebrospinal fluid with high performance liquid chromatography using electrochemical detection. *Analytical Biochemistry*, **131**, 246–53.

Soubrie, P. (1986). Reconciling the role of central serotonin neurons in human and animal behavior. *Behavioral Brain Science*, **9**, 319–64.

Spitzer, R. and Endicott, J. (1978). *Schedule of affective disorders and schizophrenia—lifetime version*, (3rd edn). New York State Psychiatric Institute, New York.

Stanley, M., Traskman-Benz, L., and Dorovini-Zis, K. (1985). Correlations between aminergic metabolites simultaneously obtained from human CSF and brain. *Life Science*, **37**, 1279–86.

Virkkunen, M. (1986). Reactive hypoglycemic tendency among habitually violent offenders. *Nutrition Review*, **44** (Suppl.), 94–103.

Virkkunen, M., Nuutila, A., Goodwin, F.K., and Linnoila, M. (1987). Cerebrospinal fluid monoamine metabolites in male arsonists. *Archives of General Psychiatry*, **44**, 241–7.

Virkkunen, M., DeJong, J., Bartko, J., Goodwin, F.K., and Linnoila, M. (1989*a*). Relationship of psychobiological variables to recidivism in violent offenders and impulsive fire setters: a follow up study. *Archives of General Psychiatry*, **46**, 600–3.

Virkkunen, M., DeJong, J., Bartko, J., and Linnoila, M. (1989*b*). Psychobiological concomitants of history of suicide attempts among violent offenders and impulsive fire setters. *Archives of General Psychiatry*, **46**, 604–6.

Wehr, T.A., Wirz-Justice, A., Goodwin, F.K., Breitmeir, J., and Craig, C. (1982). 48 hours sleep–wake cycles in manic-depressive illness: naturalistic observations and sleep deprivation experiments. *Archives of General Psychiatry*, **39**, 559–65.

Yamamoto, H., Nagai, K., and Nagakava, H. (1984). Additional evidence that the suprachiasmatic nucleus is the center for regulation of insulin secretion and glucose homeostasis. *Brain Research*, **304**, 237–41.

12

Obsessive–compulsive disorder and depression

JUAN J. LÓPEZ-IBOR, JR, JERÓNIMO SAIZ, and
ROSA VIÑAS

INTRODUCTION

The relation between obsessive disorders and mood disorders is an old concern of psychopathology. Krafft-Ebing emphasized the identity of obsessive and melancholic thoughts, and Kraepelin, Maudsley and Marchand, considered obsessions to be an integral part, or in close relationship with, the mood spectrum, specially with melancholic episodes.

Nowadays obsessive–compulsive disorders (OCD) are classified within the class of anxiety disorders, following Freud's psychodynamic approach, such as in the first edition of the Statistic and Diagnostical Manual of the American Psychiatry Association (DSM-I) (APA 1952), in the DSM-II (APA 1969), and in the ninth edition of the International Classification of Diseases of the World Health Organization (ICD-9) (WHO 1987).

However, certain clinical observations made during the 60s suggested for the first time the possibility that this supposedly unitary nosological category may be in reality be built up by entities very different from the clinical and aetiological standpoints. The DSM-III (APA 1980) proposes a reorganization of the anxiety disorders of a radically different approach. First, the term neurosis is abandoned because it implies the fact of sharing a common cause: an unconscious conflict leading to an inadequate use of the defence mechanisms. Instead of a supposed aetiology, DSM-III bases classification on the presence of common symptoms. A multiaxial system is also established to evaluate separately personality disorders, possible organic aetiologies, environmental stresses, and adaptation levels prior to the onset of the clinical manifestations. As a consequence a compulsive personality disorder is present in axis II, clearly separated from the obsessions and compulsions which characterize obsessive-compulsive disorder in axis I.

One of the most controversial aspects of the anxiety disorders in general, is the autonomy or independence from mood disorders. This has lead to two contrasting points of view, that of the defendants of a clear categorial separation (Roth et al. 1972) and that of the supporters of a continuum (unitary hypothesis, mood or affective spectrum) of emotional and affective disorders. For instance, Lewis (1934) found high rates of panic attacks and OCD in melancholic patients, which led him to a unitary vision of affective and anxiety disorders. Even before, Janet (1903) supported the notion of a clinical unity of forced agitations, a family of disorders which included obsessive disorder, panic disorder, phobic disorders, bulimia, pain syndromes, including cases of migraine and atypical facial pain, and other physiological disorders, such as the irritable colon. Janet found out that forced agitations appeared frequently associated themselves to dysthymia and depression (*psychasthenie* and *neurasthenie* respectively). The concept of depressive or affective equivalents, which includes the different forms of depressive disorders, several of the psychosomatic disturbances and many behavioural problems (López-Ibor, Jr 1972, 1991), belongs to this model.

A terminological confusion cannot be avoided in this field. Neurotic (anxiety) disorders are often referred to as emotional disorders, and this term may also include vegetative symptoms, stress reactions, and so called psychosomatic or psychophysiological conditions. On the other hand, depressive and manic disorders are called affective disorders in DSM-III. The question is, what is the difference between emotion and affect. More recently the term mood is gaining acceptance. Both in DSM-III-R and ICD-10 the class is called mood (affective) disorders. The term mood is much more correct from the psychological and psychopathological point of view (López-Ibor, Jr 1991). But, on the other hand, the mood spectrum often includes in itself the emotional and affective disorders, not a bad approach, and to make life even more uneasy, the philosopher Kant considered that mental illnesses (*Geisteskrankheiten*) were mood illnesses (*Gemütskrankheiten*) (see Grebe 1983).

López Ibor (1950, 1966) has provided a different approach. On the one side he described 'vital anxiety' as the core symptom of neurotic disorders, which was an anxiety of a biological (endothymic) nature. Vital anxiety was analogous to the vital sadness described by Schneider (1950) in depression. Nevertheless, neurotic disorders were not to be confused with other mood disorders, as there was only one of their type or subcategory. Also, López Ibor research was more concerned with aetiopathogenesis and psychopathology than with nosology. He also proposed biological treatments for neurotic conditions.

In order to make the diagnosis of OCD, DSM-III requires that obsessions and compulsions are not secondary to another mental disorder, like, for example Gilles de la Tourette's syndrome, schizophrenia, major depression, or an organic mental disorder. However, in the revised version, DSM-III-R (APA 1987) this exclusion criteria is eliminated, because it is recognized that OCD can appear together with other disorders. This is clear when OCD precedes the diagnosis of any other psychiatric disorder, such as bipolar disorder (Baer *et al.* 1985) or major depression.

Several authors propose the sentence 'with psychotic symptoms' for severe cases of OCD, the same as in mood (affective) disorders, where intrusive thoughts acquire the quality of delusion, to which the patients yield instead of showing resistance to them.

In the ICD-10 draft, OCD is considered separated from other neurotic disorders, something which is in concordance with recent investigations, that is to say, it considers OCD as an isolated disorder. But at the same time it should be remembered that OCD may be an heterogeneous condition, with different stages in its evolution and maybe clinical subforms. The different stages in the course of OCD were first described by Legrand du Saulle (1875), and well recognized by Janet (1903). The symptomatic approach to modern classifications should not make us forget that very often it is the course and not the isolated symptoms which define a disorder.

In the following we will review the data that support and the data that contradict the unique and heterogeneous character of OCD in relationship with depressive disorders. We will group this review in three parts: clinical manifestations, response to treatment, and biological research data.

CLINICAL MANIFESTATIONS

There is a considerable amount of data on the possible association between OCD and mood disorders (Insel *et al.* 1982*a*, 1984), therefore, it is quite natural that the relation between obsessions, compulsions, and depressions has been the theme of several investigations (Gittelson 1966; Videbach 1975; Welner *et al.* 1976). This topic can be considered from different perspectives.

Depressive symptoms in obsessive–compulsive disorders

It is not a rare phenomenon for OCD patients to present simultaneously affective symptoms (Jenike 1981, 1991*a*; Jenike *et al.* 1990). Rasmussen and Tsuang (1986*a*) in a sample of 100 patients diagnosed of OCD, found that 20 per cent denied suffering depression at the beginning of the study and a majority stated that depression had developed after the appearance of obsessive–compulsive symptoms. A minority presented a coincidental beginning of depression and obsessive symptoms. According to Rosenberg (1968) 34 per cent of patients with OCD had received treatment for depression some time in the past. During a depressive episode in some OCD patients the obsession increase but in others they decrease.

Many patients suffering from OCD, only seek help when they get depressed and this may exaggerate the prevalence of depression in the clinical samples. This is what the data of Kringlen (1965) seem to show. 17 per cent of his patients had depressive symptoms at the beginning of OCD, but 42 per cent had these symptoms at the moment of hospitalization. In our samples, OCD patients had a mean score for depression, using the scale of Montgomery and Asberg (1979) of 26.29 (SD = 8.67), which is considered as moderate (Davidson *et al.* 1986). Severe depression in patients with OCD, worsened the prognosis and it has been said to inter-fere with behavioural treatments (Foa 1979; Marks 1981).

Obsessions and compulsions in depressive disorders

Obsessions and compulsions are frequent in primary mood disorders (Zohar and Insel 1987*a*). In fact, it has sometimes been claimed that depressions with obsessive–compulsive symptoms represent a subtype of the mood disorder.

Obsessive symptoms were present in 14 (23 per cent) out of 61 melan-cholic patients described by Lewis (1936). Obsessive symptoms may also appear during a clearly defined major depressive disorder, usually related to the depressive symptomatology, for example obsessive thoughts of doubt.

It has been described that obsessions reduce the risk of suicide in depression (Jenike 1991*b*). In a retrospective study of 398 depressive psychotic patients, Gittleson (1966) found 152 (38.2 per cent) patients with obsessive symptoms. Among those patients, the presence of obsessions was associated with previous depressive personality traits and obsessive personality. This author also found that depressive patients with obsessive features were less likely to attempt suicide than those without those features, although they did not differ regarding the presence of feelings of despair or delusional ideas. However, Videbach (1975) could not verify this protecting effect in suicide, and Videbach (1975) and Vaughan (1976) both found out that the obsessive features were more frequent in agitated depressions than in inhibited depressions.

In a prospective study of 92 inpatients suffering from a depressive disorder, Kendall and Discipio (1970) found out that 20 (21.7 per cent) had obsessive–compulsive symptoms during the initial clinical evaluation, and that, as a group, depressive patients had very high scores in the Leyton Inventory for Obsessions (Cooper 1970) and the score increased as the illness worsened.

The course of obsessions and compulsions in depression seems to be parallel to the primary disorder. Obsessive symptoms tend to develop during depression and to decrease in proportion to the improvement of the patient; if obsessive symptoms are present before, they worsen during depression and revert to their original state the moment depression ceases (Videbach 1975; Marks 1987).

Similarities and differences of the symptoms of obsessions and depressions

Most of the difficulties in separating OCD from depression are due in both disorders to the overlapping of the symptoms of both disorders, particularly those of guilt, anxiety, doubt, and low self-esteem. Vegetative symptoms, such as weight loss and sleep disturbances, can also develop in OCD; and rituals related to food intake and sleep are frequent during a depressive episode. It is also difficult to differentiate between the brooding and morbid worries of the depressed personality from the obsessive thoughts found in OCD. Generally the brooding thoughts are egosyntonic, they are accepted by the patient as rational, perhaps exaggerated but related to the experience of the depression. On the other hand, obsessions are egodystonic, and experienced as imposed, out of context, untimely, and often resisted. While the depressed patients tend to focus on passed events, obsessive patients tend to focus on preventing future happenings.

Aggressiveness is a common phenomenon in OCD and in depressive disorders, but it is of a different nature. Depressed patients show a constantly poor control of their aggressiveness, specially during the earlier and final periods of an episode (Ledesma 1977) while OCD patients are good controllers of the same. Only very rarely do they lose this control but when it happens it leads to very extreme outbursts of self- or other-directed harmful acts.

In our OCD cases, those patients with obsessions of aggression had more disturbances of the circadian rhythm with early morning awakening. At the same time, those who were depressed and presented early morning awakening, had an episodic course. This suggests that obsessions with an aggressive content are related to depression (Viñas 1991).

Depressive and obsessive patients cope in different ways with social norms. At the turn of the century it was common in Europe to divide sets of disorders into those characterized by poor control of impulsivity on the one side and those with good control on the other. OCD belonged to the latter and moral insanity or impulsive madness (Jaspers 1909) to the other. Durkheim (1951) was the first to describe that individuals who committed suicide were not adapted to prevailing social norms, which he called anomia. On the other hand, patients who develop a depressive episode are characterized by a premorbid personality which in terms of social psychology has been named hypernomie (Kraus 1977, 1980). In other words, vulnerability to depression is characterized by an excessive adaption to social norms. When compared with this cases, OCD patients present a different picture. On the one hand they seem to adapt extremely well to norms, as they are perfectionistic. But on the other, this adaptation only reflects a particular way of coping which at the end becomes nonadaptative (unsignificant substitutes for the significant) (Janet 1906). Therefore, once the activity becomes ritualistic, it loses its sense and as a consequence, washing rituals do not result in cleanliness, moral obsessions make the patient behave in the most immoral way as he becomes the source of his moral principles, and order as an end challenges the efficiency of keeping order as a means for something else.

Clinical course

The natural course of both illnesses is different. In contrast to depression which tends to be episodic OCD is usually chronic (Coryell 1981). The age of onset OCD is lower in comparison with depression and the male/female ratio is also different (lower for depression).

Kendell and Discipio (1970) observed that obsessive symptoms were rare in mania. The disappearance of obsessive–compulsive symptoms has also been described in OCD and bipolar disorders during manic episodes (Gordon and Rasmussen 1988). Similarly, in the personality profile of our sample according to the Mini-Mult (Kinkannon 1968), there is a predominance in the scales of psychasthenia and depression, and the lowest score corresponded to hypomania (Viñas 1991).

Another minor but interesting aspect is the fact that depressive symptoms get better during pregnancy while the contrary happens in OCD (Brandt and Mackenzie 1987).

The association with motor disorders and neurological findings

A very important factor which differentiates between depressive disorders and OCD is the association of the latter with motor disorders, specially with Gilles de la Tourette's syndrome, but also with disorders such as Sydenham's chorea (Swedo *et al.* 1989). This fact has been correlated with the possible involvement of dopaminergic disturbances in OCD and the response to neuroleptics or the so-called treatment augmentation with neuroleptics.

Family studies

Depression and OCD tend to appear in the same families, suggesting an association between both illnesses, although the nature of this association is not clear. In a retrospective study, Coryell (1981) has found out that obsessive patients have, with greater frequency, a family history of mania or depression when compared to a control group of patients with a non-affective primary disorder (4.6 per cent versus 20.9 per cent). In our samples, there is a high percentage of a positive family history of mood disorders: major depression (12.9 per cent) and dysthymic disorders (12.9 per cent).

Epidemiological surveys

There is also an association of both syndromes in the general population. Boyd *et al.* (1984) found that, using a multiple diagnosis system, patients with OCD had a probability of suffering depression 10.8 times greater than individuals without OCD.

The symptoms of OCD and of depression often appear at the same time in the same patient. In the USA, the ECA study has found that approximately one-third of the patients diagnosed as having OCD also fulfilled DSM-III-R diagnosis criteria for major depression. Therefore, it seems that depression is the most frequent complication of OCD. It can develop before, during, or after the obsessions (Marks *et al.* 1975), and both syndromes can vary at the same time or be independent in their course. Furthermore a high percentage of patients with OCD, develop a serious depression along their evolution.

Secondary depression in obsessive–compulsive disorders. Comorbidity between both disorders

Rasmussen and Tsuang (1986*b*) in a sample of 44 patients found that 80 per cent presented dysphoric symptoms and that 75 per cent fulfilled criteria for past or present major depression, and Rasmussen and Eisen (1988) found an incidence of depression over 67 per cent, similar to that found in panic disorders. Noshirvarni *et al.* (1991) found that 14 per cent of the patients had a secondary depression and 11 per cent dysthymic disorders according to Research Diagnostic Criteria.

In many patients with OCD, however, depressive episodes survive after a long evolution of obsessive disorder. This has led to attempts to differentiate the diagnosis of OCD and depression on the basis of which precedes which. That is to say, a diagnosis of OCD can only be made if the depressive symptoms appeared later in the evolution. According to this diagnosis scheme, it is presumed that the primary disorder is the basic one and the secondary is the one derived from it.

Although longitudinal studies show that depression is a frequent complication of OCD, many OCD patients will never suffer a depression. In a retrospective study Welner *et al.* (1976) found that the transition of obsessions to depressions was present three times more frequently (38 per cent of the cases) than the transition of depressions to obsessions (11 per cent).

When depressive symptoms preceded obsessions, the prognosis was better. The chronological sequence of depressions and obsessions seems to have as much theoretical importance as practical clinical utility.

The subcategory of good prognosis obsessive–compulsive disorder

On the other hand, several classifications of obsessive disorders do isolate a subgroup related to the mood disorders (Costa Molinari 1971) and Insel (1982) has described an affective subgroup with a late onset and a good prognosis.

From all these studies it can be considered that OCD and depressive disorder are different conditions but that obsessive symptoms are common in depression and depressive symptoms in OCD. The nature of this relationship is not clear. Obsessive symptoms seem to add little to a depressive disorder but depressive symptoms in OCD may play a more significant role. Sometimes they may be considered as secondary to the OCD but at other times they may differentiate a subtype related to mood disorders (late onset, predominantly in females, with more depressive symptoms in the course of or present illness, with obsessions of aggression, a more episodic course, and a better prognosis and response to treatment), from another more autoctonous one. More research is needed to clarify these aspects.

RESPONSE TO TREATMENT

The response to treatment has been used as evidence in favour of the mood spectrum model. On the other hand, the concept of masked depression has been too widely used in many countries, through the influence of the promotion of antidepressants in general practice (López-Ibor, Jr 1991*a*). Nevertheless, based on this approach the response to a pharmacological treatment can be used to identify disorder groups sharing a common pathophysiology. Eight disorders: depression, bulimia, panic disorders, OCD, attention deficit disorder with irritability, cataplexia, migraine, and irritable colon syndrome, have been named as 'disorders of the affective spectrum' by Hudson and Pope (1990), and the idea of a group of serotonin system related disorder or SSRD, is the subject of great interest and discussion today (López-Ibor, Jr 1988).

Some tricyclic antidepressants and MAOI have demonstrated their efficacy in OCD. This overlapping in the response to the treatment is quoted as evidence that OCD is biologically related to endogenous depression. Nevertheless, OCD symptoms do not seem to respond to electroconvulsive therapy as do the symptoms of endogenous depression (Lieberman 1984).

It can also be considered that OCD, like other nonmood (affective) disorders (panic disorder, other anxiety disorders, bulimia, enuresis, migraine, and chronic pain syndrome) responds to tricyclic antidepressants even when there is no initial primary or secondary depression (Murphy *et al*. 1985), and therefore this can be used only as an approximation to the classification of the disorders of the mood spectrum. Furthermore, the

TABLE 12.1. *Clomipramine in obsessive–compulsive disorders*

Reference	Type of study	Comments
López-Ibor *et al.* 1967, 1969	Open	Positive
Jiménez Garcia 1967	Open	Positive
Laboucarié *et al.* 1967	+ECT, open	Positive
Renynghe de Voxrie 1968	Open	Positive
Marshal 1971	Open	Positive
Waxman 1973, 1975	Open	Positive
Solyom and Sookman 1977	Open	Positive
Wyndowe *et al.* 1975	Open	Positive
Yaryura-Tobias *et al.* 1976	DB placebo	Positive
Karabanow 1977	DB placebo	Positive
Waxman 1977	DB diazepam	Positive
Greingras and Bhagavan 1977	Open	Positive
Yaryura-Tobias 1977	Tryptophan	Positive
Ananth *et al.* 1979	Open	Positive
Ananth *et al.* 1979	DB amitryptiline	Positive
Montgomery 1980	DB cross over, placebo	Positive
Thoren *et al.* 1980*a*	DB nortriptiline/placebo	
Jaskary 1980	DB mianserine	No diff.
Rapoport *et al.* 1980	DB desipramine placebo	No diff.
López-Ibor *et al.* 1973	Open, CMI+5-HTp	Positive
Cassano *et al.* 1981	DB haloperidol/diazepam/placebo	
Insel *et al.* 1983	DB clorgiline/placebo	Positive
Mavissakalian *et al.* 1983, 1985	DB imipramine/placebo	Positive
Rasmussen 1984	Open, lithium/triptophan	Positive
Volankas *et al.* 1985	DB imipramine	No diff.
Warneke 1985	Open	Positive
Flament *et al.* 1985	DB cross over, placebo	Positive
Lipsedge *et al.* 1987	Open, clonidine	Positive
Jenike *et al.* 1989*a*	DB placebo	Positive
Leonard *et al.* 1989	DB cross over desipramine	Positive
De Veaugh-Geiss *et al.* 1989	DB placebo	Positive
Greist *et al.* 1990	DB placebo	Positive
Pato *et al.* 1991	DB buspirone	No diff.

DB = double blind, randomized.

finer differences between different disorders can be seen by looking at the data in more depth, doing a so-called pharmacological dissection (Klein 1989), as each one of these disorders responds better to one type of antidepressants than to others. OCD is similar to depression in that the same drugs that have been found to be effective for OCD are drugs that have also proved to be effective in depressions. But, on the other hand, OCD responds only to drugs which act on serotonin metabolism and not to the rest of antidepressants, which makes OCD a more specific illness than depression (Montgomery *et al.* 1990).

The fact that only the inhibitors in the serotonin reuptake are effective in OCD, while the inhibitors in noradreline reuptake are not, has no parallel with mood disorders nor with other anxiety disorders. This observation was originally made more than 20 years ago for clomipramine by López-Ibor, Jr *et al.* (1967, 1969) and later corroborated in many studies (Table 12.1). This leads to the hypothesis that some kind of serotonergic dysfunction is present in OCD. The success of the serotonergic hypothesis emerges from studies which show the efficacy of the new selective serotonin reuptake inhibitors (SSRIs). Fluvoxamine, fluoxetine, and in a minor way, mianserine, as well as other nontricyclic antidepressants have shown antiobsessive effects in clinical studies (Table 12.2). This fact is supported also by the findings that, in the case of a poor response, strategies oriented to enhance the serotonergic potency of the treatment, the so-called treatment augmentation strategies, seem to work in clinical conditions.

Depressive symptoms in OCD seem to respond in a parallel way to improvement of obsessive and compulsive symptoms (Flament *et al.* 1985; Perse *et al.* 1987; Goodman *et al.* 1989*a*; Cottraux *et al.* 1990), although older studies do not show this parallelism between depressive and obsessive symptoms (Ananth *et al.* 1981; Insel *et al.* 1983). It is interesting that in the study of Goodman *et al.* (1989*a*), the response of depressive symptoms was in contrast to the one seen in depression; fast, with significant differences when compared to placebo after two weeks treatment. This timing in the response is different to the one observed in the treatment of depression and seems to show that the depressive symptoms form an integral part of OCD.

Some studies show an antiobsessive effect of the clomipramine independently of their antidepressive effect (Flament *et al.* 1985; Mavissakalian *et al.* 1985; Murphy *et al.* 1985; Montgomery and Fineberg, 1987; Goodman *et al.* 1990). Pulman *et al.* (1984) proved the efficacy of clomipramine in patients with repetitive behaviour, which did not form part of the classical obsessive neurosis, which would not support this antiobsessive and/or anticompulsive effect. The study of Marks *et al.* (1980), the review of Marks (1983) and the one of Mavissakalian and Michelson (1983) are exceptions which seem to support Marks' hypothesis with regard to the

TABLE 12.2. *Serotonin reuptake inhibitors in OCD*

Reference	Type of study	Comments
Prasad 1985	Open, trazodone	Positive
Prasad 1984	DB zimeldine/ imipramine	ZMB > IMI response?
Kahn 1984	Zimeldine	
Fontaine and Chouinard 1985	Open, zimeldine	
Pato *et al.* 1991	DB CMI/buspirone	Both positive
Insel *et al.* 1985	DB zimeldine, demethyl- imipramine, CMI, placebo	DMI = ZMD CMI > DMI/ZMD
Price *et al.* 1987	Single-blind fluvoxamine/ placebo	Positive
Perse 1987	DB cross over fluvoxamine/placebo	Positive*
Goodman *et al.* 1989*a,c*	DB, fluvoxamine/ placebo	Positive*
Goodman *et al.* 1989*b*	DB fluvoxamine/ demethylimipramine	Fluvox > DMI*
Turner *et al.* 1985	Single blind, fluoxetine	Positive†
Fontaine and Chouinard 1986	Open, fluoxetine	Positive
Jenike *et al.* 1989*b*	Open, fluoxetine	Positive*
Levine *et al.* 1989	Fluox	
Liebowitz 1991	Fluox	
Riddle 1990	Fluox	
Pigott *et al.* 1990	DB CMI/fluoxetine	No. diff.
Jenike *et al.* 1990	Metanalysis of open studies, fluoxetine	Fluox better tolerated
Blick and Hackett 1989	DB, sertraline/placebo nondepressed patients	

* Response not related to baseline depressions.
† Effect on depression and rituals.
DB = double blind; CMI − clomipramine.

particular usefulness of antidepressants in the presence of marked mood disturbances. What is possibly demonstrated in the Marks' study (1980), which includes a behaviour therapy, it is not that clomipramine is ineffective in nondepressed patients with OCD but that the behaviour therapy alone is ineffective in depressed OCD patients (Liebowitz *et al.* 1988).

Another fact that differentiates OCD from depression is the poor response to placebo observed in OCD (Mavissakalian *et al.* 1990*a*). For example, in the study of Montgomery (1980), with placebo the response was present in 5 per cent of the patients, while 65 per cent of the patients responded to clomipramine. In the multicentre study of De Veaugh-Geiss *et al.* (1989), the placebo response was also very low and the benefit of clomipramine was confirmed. On the other hand in depression there was a response to placebo of 30 per cent or even more, up to 50 per cent can be expected. Furthermore, the depressive symptoms present in OCD do not seem to respond to placebo. Besides, it seems that depressive symptoms also do not respond to drugs which do not respond to OCD. Any effect of a nonserotonergic drug is weak. The general lack of response to depressive and obsessive–compulsive symptoms to placebo or to the mentioned nonserotonergic drugs, suggests that depressive symptoms form an integral part of OCD and respond as a part of the treatment of OCD. This aspect differentiates OCD markedly from depression and therefore it seems that the mechanism of action of the selective serotonin reuptake inhibitors requires an adaptation period of the receptors from the antidepressive effect to take place, which is different in the case of depression forming part of OCD, where the effect would be more direct.

Sleep deprivation generally produces a brief antidepressive response in half of the depressed patients (Joffe and Brown 1984). It has been suggested that the positive response predicts a positive response to anti-depressants treatment (Roy-Byrne *et al.* 1984). Patients suffering from OCD do not respond in a global way to sleep deprivation, and their obsessions and/or compulsions remain unchanged (Joffe and Swinson 1988; Swinson and Joffe 1988).

Thoren *et al.* (1980*b*), have found a correlation between a reduction of the concentration of 5-HIAA, the serotonin metabolite, in cerebrospinal fluid (CSF) and the improvement of the obsessive-compulsive disorders during treatment with clomipramine. From the therapeutic point of view what seems to be important is the reduction of 5-HIAA concentration, but not of 4-hydroxy-3-methoxy-phenylglycol (MHPG), the main metabolite of the catecholamines, as there is a significant correlation between the reduction of 5-HIAA concentration and the therapeutic effect on the rituals and obsessive thinking, this effect being more pronounced the greater the reduction of 5-HIAA. These results differed from the ones obtained on the dysphoryc symptoms which were better the greater the diminution of MHPG concentration.

In young patients, Flament *et al.* (1987) found a significant negative correlation between the serotonin platelet concentrations and the severity of the clinical symptomatology. Furthermore, during a clomipramine treat-ment a diminution of the serotonin platelet concentration was achieved, which correlated with all the parameters of clinical improvement. A high serotonin concentration in platelets before treatment was a predictor of a favourable clinical response. On the other hand the clinical severity does not predict the response to the drug. In the study of Thoren *et al.* (1980*b*) patients who responded to treatment with clomipramine had higher pre-treatment 5-HIAA CSF concentrations than those who did not respond, and the clinical improvement correlated positively with the reduction of 5-HIAA concentration during treatment. Flament *et al.* (1987) have also described a reduction of 10 per cent in the platelet MAO activity during treatment with clomipramine, and that this diminution was correlated with changes in the obsessive and depressive symptoms. This association could be interesting since it has been speculated that the activity of the platelet MAO activity reflects the central serotonergic metabolism, as the MAO activity of the platelets correlated in a significant way with the plasmatic concentration of prolactin after a serotonergic stimulation (Kleinman *et al.* 1979) and with the 5-HIAA CSF concentration in healthy volunteers (Oreland *et al.* 1981).

Pigott *et al.* (1990) have found during a double-blind controlled study with clomipramine and fluoxetine in the treatment of OCD, that the 5-HT platelet concentrations revealed significant and equivalent reductions after both treatments, with clomipramine and fluoxetine. Both drugs have proved to be useful for the treatment of OCD.

In some studies (Stern *et al.* 1980; Insel 1983; Mavissakalian *et al.* 1990*b*) the plasma concentration of clomipramine, but not of desmethyl-clomipramine, correlated significantly with the reduction of obsessive–compulsive symptoms. This could explain the greater effectiveness of clomipramine used in high doses and principally intravenously (50 per cent of the dosage) avoiding the first-pass effect through the liver and therefore the metabolization to desmethyl-clomipramine (López-Ibor, Jr 1969).

BIOLOGICAL MARKERS

While the nature of the relation, if it exists, between OCD and depression remains unsolved, one way of studying it is the identification of biological markers which both disorders have in common. Nevertheless, although biological disturbances common to both disorders, depression and OCD, have been identified, the specific nature of this relation is still unclear.

Dexamethasone suppression test

The dexamethasone suppression test (DST) has been proposed as a specific laboratory test for the diagnosis of melancholia (Carroll *et al.* 1981). This test identifies an alteration in the hypothalmic–hypophyseal–adrenal axis. Lieberman *et al.* (1985) did not find an OCD patient who was a non-suppressor, compared to the 37 per cent of nonsuppressor in the depressed patients group. Insel *et al.* (1982*a*) have found that 37.5 per cent of OCD patients with higher scores in the Hamilton Depression Rating Scale (HDRS) and a greater familiar history of mood disorders were non-suppressors. In a later study, they found a proportion of 25 per cent and again, only in those with higher HDRS scores, when compared to the suppressors (Insel *et al.* 1984). Jenike (1991*a*) found that 17 per cent of an OCD patient sample were nonsuppressors, but, four out of five non-suppressing patients were clinically depressed. Monteiro *et al.* (1986) obtained a rate of 4 per cent and Cottraux *et al.* (1989) of 30 per cent. Recently, Coryell *et al.* (1989) did also not find any nonsuppressors among nondepressed OCD patients, the proportion being that of the healthy controls. Cottraux *et al.* (1984, 1989) found they were present at a rate of 20 per cent and in the light of these results they consider DST as a possible depression marker in OCD patients. Vallejo *et al.* (1989) also consider that non-suppression depends on the concomitant presence of OCD in major depression.

Sleep architecture

The architecture of sleep, is a widely used biological marker of depressions. OCD patients tend to have a shortening of the REM sleep latency, a reduction of the delta sleep, a reduction of the total sleep, and a reduction of the efficacy of sleep. Curiously the REM density, which is increased in primary depressions (even in remission), is normal in OCD patients. Therefore there is an overlapping in the physiology of sleep between both disorders, but there are also subtle differences (Insel *et al.* 1982*b*, 1984). Rapoport *et al.* (1981) have described anomalies in the sleep of adolescents with OCD which were similar to the ones found in depressed patients of medium age.

Central nervous system abnormalities

Following the association of OCD with motor symptoms, central nervous system abnormalities have been investigated in OCD, and here again the evidence is in favour of the difference with depressive disorders. The following have been identified: presence of soft neurological signs (Conde Lopez *et al.* 1990*a,b*; Hollander *et al.* 1990), abnormal images in MRI (Garber *et al.* 1989), abnormal PET scanning (Baxter *et al.* 1987, 1988, 1990), abnormal electric activity in brain mapping (Malloy *et al.* 1989; Towey *et al.* 1990), abnormal CT-scan images (Luxemberg 1988), and even catatonic symptoms (Hermesh *et al.* 1989).

Growth hormone secretion postclonidine

An other biological marker of the depression disorders is the modification of the growth hormone (GH) plasma concentration after the administration of clonidine. This response is blunted both in depressed and in OCD patients, independently if the latter have or have not a secondary depression (Siever *et al.* 1983; Insel *et al.* 1984).

Neuroendocrine response to serotonergic probes

One of the most fruitful approaches for the study of the central serotonergic function, is the use of the so-called neuroendocrine probes. They measure the modifications in the blood concentration of hormones, whose secretion is controlled by hypothalmic factors which themselves are under the control of serotonin. They are used to evaluate in an indirect form, the integrity of the functioning of the central serotonergic tract.

We have undertaken several studies with serotonergic probes on the different clinical subgroups of depressive disorders, included in the so-called 'mood spectrum' (López-Ibor, Jr *et al.* 1988), from anxiety disorders to major depression with melancholia. In depressive disorders we have found a negative correlation between the increase of prolactin concentration after a serotonergic challenge with fenfluramine or intravenous clomipramine, and the greater severity of the disorder, or by the fact of being endogenous according to the Newcastle scale which also correlates negatively with the increase in the prolactin secretion. Most of the research data in this area points out the presence in depression of a blunted prolactin and GH response to the serotonergic stimulus and that this alteration is mostly present in depressions classically considered to be more endogenous or severe (Fig. 12.1) or more melancholic (Fig. 12.2). Other serotonergic challenges have produced similar results, such as a blunted response of prolactin after tryptophan administration in depressed patients (Price *et al.* 1987).

On the other hand, findings in OCD are quite different. Charney *et al.* (1988) found increased plasma prolactin responses in OCD, similar to those of controls. Fineberg *et al.* (1991) also observed significant increases of prolactin, GH, and plasma concentrations after stimulation with L-tryptophan.

In a similar way, Bastani *et al.* (1990), studied the cortisol and prolactin plasma concentrations, and prolactin, and the behavioural response to the oral administration of MK-212 (6-chlor-2-(1-piperanizil)-pyperazine), a serotonin agonist and placebo in 17 patients suffering from OCD and nine controls. Baseline prolactin and cortisol plasma concentration did not differ in a significant way in the two groups. But, the response to the oral

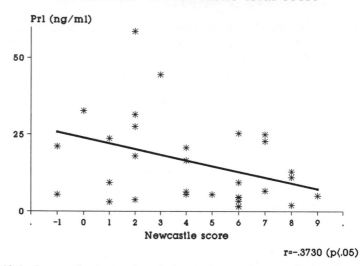

Prl increase vs Newcastle total score

r=-.3730 (p⟨.05)

FIG. 12.1. In a mixed sample of depressive patients suffering from major depression with melancholia, major depression, and dysthymic disorder, according to DSM-III criteria, the maximum increase in prolactin concentration after fenfluramine administration correlates negatively with the score on the Newcastle scale which measures endogeneity.

administration of 20 mg of MK-212, was significantly blunted in patients with OCD, compared with the results of the controls, X, for prolactin and cortisol.

From the a behavioural point of view, Zohar *et al.* (1987*c*), using the serotonergic agonist of the postsynaptic receptor, *m*-CPP (Mueller *et al.* 1985) have found that a single oral dose will produce a transitory but marked exacerbation of the obsessive–compulsive symptoms, even with the appearance of new obsessions, which does not occur in controls nor in patients with OCD after placebo. *m*-CPP, has almost the same affinity for the 5-HT_{1A}, 5-HT_{1D}, and 5-HT_2 and a more powerful affinity for 5-HT_{1C} (Khanna 1991) receptor for alpha-2 adrenergic sites in homogenized human cerebral tissue (Hamik and Peroutka 1989). Metergolin, a serotonin-receptor blocker, produces a reduction of this response (Murphy 1988). But, Charney *et al.* (1988) with a similar design did not find significant differences between controls and OCD patients in the behavioural response to intravenous administration of *m*-CPP (0.1 mg/kg) or L-tryptophan (7 g); and no global differences in the prolactin response to *m*-CPP or to tryptophan, nor differences between OCD patients and controls in the response of cortisol to *m*-CPP. Furthermore, MK-212 does not

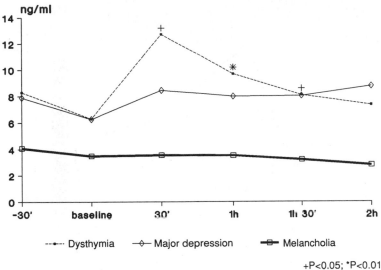

Prolactin post CMI in depressions

--•-- Dysthymia   --◆-- Major depression   --■-- Melancholia

+P<0.05; *P<0.01

FIG. 12.2. Prolactin concentration after clomipramine: patients with major depression with melancholia have lower baseline concentrations and a blunted response to the challenge, while dysthmic patients have a more normal response.

produce any change in the intensity of the obsessive–compulsive symptoms (Bastani *et al.* 1990).

We have carried out studies using clomipramine as a challenge (12.5 mg in a 10 min intravenous infusion) and have found an increased hormonal response to the serotonergic stimulus in OCD patients. The differences were significant for the maximun increases of prolactin when compared to controls. For cortisol, the increase was significant 30 min and 120 min after clomipramine infusion. For GH, the differences were significant in all samples (Figs 12.3, 12.4, and 12.5). We had excluded study patients who fulfilled DSM-III-R criteria for major depression. Nevertheless, some of them had a significant, but subdiagnostic amount of depressive symptoms (secondary depression), which were evaluated with the MADRS. The maximum difference of prolactin secretion correlated negatively with the severity of depression evaluated with the Montgomery and Asberg scale (1979) (Fig. 12.6), and that this negative correlation was also present for the items of manifested sadness and suicidalness. Blunted responses to clomipramine also correlated with the circadian mood disturbance, early awakening, and suicidalness items of the AMDP system (Bobon 1983) (Table 12.3). Furthermore, a subgroup of eight patients which presented a

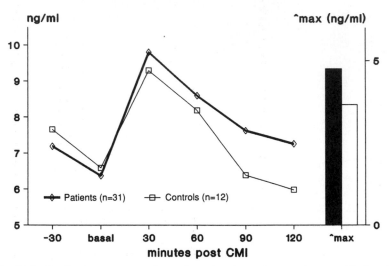

FIG. 12.3. Prolactin concentration after clomipramine: patients suffering from obsessive–compulsive disorder have a similar response to an age- and sex-matched control group.

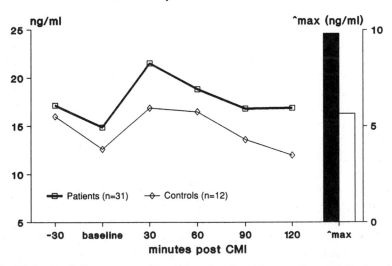

FIG. 12.4. Cortisol concentration after clomipramine: patients suffering from obsessive–compulsive disorder have a higher response than an age- and sex-matched control group. The differences are not statistically significant.

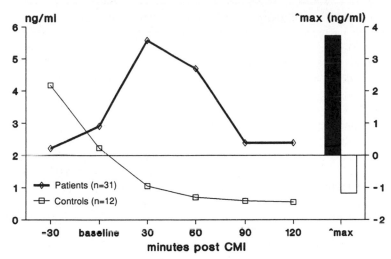

FIG. 12.5. Growth hormone concentration after clomipramine: patients suffering from obsessive–compulsive disorder have a higher response than an age- and sex-matched control group.

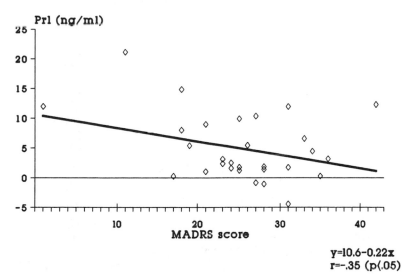

FIG. 12.6. The score in the Montgomery–Äsberg Depression Rating scale in a sample of obsessive–compulsive patients not fulfilling criteria for major depression, correlates negatively with the maximum increase in the prolactin secretion.

TABLE 12.3. *Psychopathology and prolactin secretion postclomipramine.*

	Baseline	30	60	90	120	Max
MADRS (r)						
Total score	ns	ns	ns	ns	ns	r: = −0.35; $p < 0.05$
Manifest sadness	ns	ns	ns	ns	ns	r: = −0.35; $p < 0.05$
Suicidal thoughts	ns	ns	ns	ns	ns	r: = −0.41; $p < 0.05$
AMDP (MW)						
Obsessive impulses	ns	$p < 0.05$	ns	ns	ns	ns
Circadian mood disturbances	ns	$p < 0.05$	ns	ns	ns	$p < 0.01$
Suicidalness	ns	$p < 0.1$	ns	ns	ns	$p < 0.05$
Early awakening	ns	ns	ns	ns	ns	$p < 0.01$
Affective liability	ns	ns	ns	ns	ns	$p < 0.1$

r: correlation coefficient.
MW: Mann–Whitney for scores 0–2 versus 3–4.
(ns = not significant).

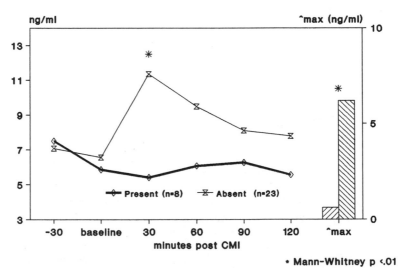

**Prolactin post 12.mg CMI in OCD
vs circadian mood disturbance (AMDP)**

• Mann-Whitney p <.01

FIG. 12.7. Prolactin concentration after clomipramine: obsessive–compulsive patients with circadian mood changes (AMDP) have a more blunted response, the same can be found in depressive disorders.

circadian mood disturbance had a blunted prolactin response to the serotonergic stimulus (Fig. 12.7). In addition, this subgroup of patients with
circadian mood disturbances had a higher baseline cortisol concentration,
as one would expect in situations of chronic stress and in depression. These
data suggest that OCD patients with symptoms of depression may differ
from those without them when confronted with a serotonergic challenge.
Therefore, we separated them into three groups: those with a score of less
than 19, those with a score of 19 to 27, and those with a score of over 27
points. The prolactin response to the clomipramine challenge was different
in each of the groups, being very blunted in those with higher scores, and
the difference was statistically significant when comparing the two extreme
groups (Fig. 12.8).

We have also divided our sample according to the type of obsession and
compulsion present, using the Y-BOCS scores. Here, the only significant
difference which emerges is related to the presence of compulsions of hand
or body washing. The subgroup of the patients with these type of symptoms have, when compared to the rest, a higher prolactin response to
clomipramine (Fig. 12.9), as if they represented a purer form of OCD.

Our results suggest that depression can be associated with OCD, and

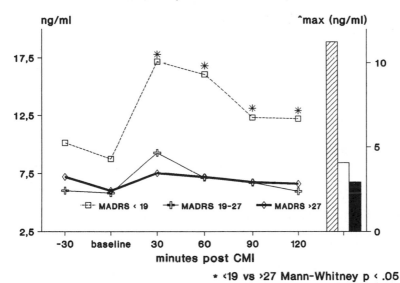

FIG. 12.8. The same sample is subdivided in three subgroups according to the
score in the Montgomery–Åsberg Depression Rating scale. Obsessive–compulsive
patients with a low score, that is with very few depression symptoms, have a high
response to clomipramine in comparison to those patients with more prominent
depressive symptoms.

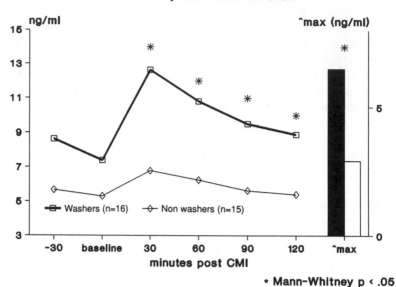

Prolactin post CMI in OCD

FIG. 12.9. The same sample is subdivided according with to the type of obsessive and compulsive symptoms. Patients presenting with washing rituals have an exaggerated response to the clomipramine administration when compared with the rest of the patients.

than when this is the case, the neurochemical changes related to the serotonin metabolism are deeply changed in the direction of becoming what could be expected in depressed patients (i.e. a blunted prolactin response). But on the other hand, the purest forms of OCD patients, those nondepressed or those with washing rituals show the reverse image, a greater response to the serotonergic challenge. In both cases, the response to treatment can be explained but through different mechanisms, one through the increase in the serotonin metabolic activity, and, in the other case, through down regulation of serotonin receptors. These would have been previously hypersensitive, as other evidence from the literature with *m*-CPP or MK-212 suggests.

CONCLUSIONS

The available clinical or neurochemical data suggest that:

1 In spite of some comorbidity, OCD and depression disorders are different clinical conditions.

2 Obsessive and compulsive symptoms belong to the realm of the symptoms of depressive disorders.
3 Depressive symptoms are present in the evolution of OCD. They take the form of secondary depression, which respond to treatment the same way as the rest of the symptoms of OCD. These depressive symptoms seem to correlate with some neurochemical findings.
4 The response to serotoninic antidepressants is not related to the presence of this secondary depression.

REFERENCES

Ananth, J., Pecknold, J.C., Van der Steen, N., and Engelsman, F. (1979). Double-blind comparative study of chlorimipramine in obsessive neurosis. *Current Therapeutic Research*, **25**, 703–9.

Ananth, J., Pecknold, J., Van Den Steen, N., and Engelsmann, F. (1981). Double-blind comparative study of clomipramine and amitriptyline in obsessive neurosis. *Progress in Neuro-Psychopharmacology*, **5**, 257–62.

APA (American Psychiatric Association) (1952). *Diagnostic and statistical manual of mental disorders (DSM-I)* (1st edn). American Psychiatric Press, Washington, DC.

APA (American Psychiatric Association) (1969). *Diagnostic and statistical manual of mental disorders (DSM-II)* (2d edn). American Psychiatric Press, Washington, DC.

APA (American Psychiatric Association) (1980). *Diagnostic and statistical manual of mental disorders (DSM-III)* (3rd edn). American Psychiatric Press, Washington, DC.

APA (American Psychiatric Association) (1987). *Diagnostic and statistical manual of mental disorders (DSM-III-R)* (3rd revised edn). American Psychiatric Press, Washington, DC.

Asberg, M., Montgomery, S.A., Perris, C., Schalling, D., and Sedvall, G.A. (1978). Comprehensive psychopathological rating scale. *Acta Psychiatrica Scandinavica*, **56** (Suppl. 271), 4–11.

Baer, L., Minichiello, W.E., and Jenike, M.A. (1985). Behavioral treatment in two cases of obsessive–compulsive disorder with concomitant bipolar affective disorder. *American Journal of Psychiatry*, **142**, 358–69.

Bastani, B., Nash, F., and Meltzer, H.Y. (1990). Prolactin and cortisol responses to MK-212, a serotonin agonist, in obsessive–compulsive disorder. *Archives of General Psychiatry*, **47**, 833–9.

Baxter, L.R., Phelps, M.E., Mazziota, J.C., Guze, B.H., Schwartz, J.M., and Selin, C.E. (1987a). Local cerebral glucose metabolic rates in obsessive–compulsive disorder: a comparison with rates in unipolar depression and in normal controls. *Archives of General Psychiatry*, **44**, 211–19.

Baxter, L.R., Thompson, J.M., Schwartz, J.M., Guze, B.H., Phelps, M.E., Mazziotta, J.C., Selin, C.E. et al. (1987b). Trazodone treatment response in obsessive–compulsive disorder correlated with shifts in glucose metabolism in the caudate nuclei. *Psychopathology*, **20** (Suppl. 1), 114–22.

Baxter, L.R., Schwartz, J.M., Mazziotta, J.C., Phelps, M.E., Pahl, J.J., Guze, B.H. *et al.* (1988). Cerebral glucose metabolic rates in nondepressed patients with obsessive–compulsive disorder. *American Journal of Psychiatry*, **145**, 1560–3.

Baxter, L.R., Jr, Schwartz, J.M., Guze, B.H., Bergamn, K., and Szuba, M.P. (1990). PET Imaging in obsessive–compulsive disorder with and without depression. *Journal of Clinical Psychiatry*, **51** (Suppl. P), 61–9; discussion 70.

Benkelfat, C.H., Murphy, D.L., Zohar, J., Groser, G., and Insel, T.R. (1989). Clomipramine in obsessive–compulsive disorder, further evidence for a serotonergic mechanism of action. *Archives of General Psychiatry*, **46**, 23–8.

Blick, P.A. and Hackett, E. (1989). Sertraline is effective in obsessive–compulsive disorder. In *Psychiatry today: VIII World Congress psychiatry abstracts: 152* (ed. C.N. Stefanis, C.R. Soldatos, and A.D. Ravabilias). Elsevier, New York.

Bobon, D.P. (ed.) (1978). *Le manual AMDP*. Liège.

Boyd, J.H., Burke, J.D., Gruenberg, E., Holzer, C.E., III, Rae, D.S., George, L.K. *et al.* (1984). Exclusion criteria of DSM-III: a study of co-occurrence of hierarchy-free syndromes. *Archives of General Psychiatry*, **41**, 938–89.

Brandt, K.R. and Mackenzie, T.B. (1987). Obsessive–compulsive disorder exacerbated during pregnancy: a case report. *International Journal of Psychiatry and Medicine*, **17**, P361–6.

Carroll, B.J., Feinberg, M., Greden, J.F., Tarika, J., Albala, A.A., Haskett, R.F. *et al.* (1981). Specific laboratory test for the diagnosis of melancholia. *Archives of General Psychiatry*, **38**, 15–22.

Cassano, G.G., Castro Giovani, P., and Mauri, M. (1981). A multicenter controlled trial in phobic obsessive psychoneurosis. The effect of chlorimipramine and its combinations with haloperidol and diazepam. *Progress in Neuropsychopharmacology*, **5**, 129–38.

Conde Lopez, V., De La Gandara Martin, J.J., Blanco Lozano, M.L., Cerezo Rodriguez, P., Martinez Roig, M., and De Dios Francos, A. (1991). Signos neurológicos menores en los trastornos obsesivo–compulsiovs. *Actas luso-españolas de Neurología, Psiquiatría y Cientias Afines*, **19**, 1–21.

Cooper, J. (1970). The Leyton obsessional inventory. *Psychiatric Medicine*, **1**, 48.

Coryell, W.H. (1981). Obsessive–compulsive disorder and primary unipolar depression. Comparison of background, family history, course and mortality. *Journal of Nervous and Mental Diseases*, **169**, 220–4.

Coryell, W.H., Black, D.W., Kelly, M.W., and Noyes, R., Jr (1989). HPA axis disturbances in obsessive–compulsive disorder. *Psychiatry Research*, **30**, 243–51.

Costa Molinari, J.M. (1971). Classificación, clínica, evolución y diagnóstico. In *XI National Congress on neuro-psychiatry*, Málaga, pp. 159–77.

Cottraux, J.A., Bouvard, M., Claustrat, B., and Juenet, C. (1984). Abnormal dexamethasone suppression test in primary obsessive–compulsive disorder patients: a confirmatory report. *Psychiatry Research*, **13**, 153–65.

Cottraux, W., Claustrat, B., Mollard, E., and Sluys, M. (1989). Depression and dexamethasone suppression test in 50 obsessive–compulsive patients. *Journal of Anxiety Disorders*, **3**, 7–13.

Cottraux, J., Mollard, E., and Bouvard, M. (1990). A controlled study of fluvoxamine and exposure in obsessive–compulsive disorder. *International Clinical Psychopharmacology*, **5**, 17–30.

Charney, D.S., Goodman, W.K., Price, L.H., Woods, S.W., Rasmussen, S.A., and Heninger, G.R. (1988). Serotonin function in obsessive–compulsive disorder: a comparison of the effects of tryptophan and *m*-clorophenyl-piperazine in patients and healthy subjects. *Archives of General Psychiatry*, **45**, 177–85.

Davidson, J., Turnbull, D., Strickland, R., Miller, R., and Graves, K. (1986). The Montgomery–Asberg depression scale: reliability and validity. *Acta Psychiatrica Scandinavica*, **73**, 544–8.

De Veaugh-Geiss, J., Landau, P., and Katz, R. (1989). Treatment of obsessive–compulsive disorder with clomipramine. *Psychiatric Annals*, **19**, 97–101.

Durkheim, E. (1951). *Suicide*. Free Press, Glencoe, IL.

Fineberg, N., Bullock, S.A., Montgomery, S.A., Da Rosa Davies, J., and Cowan, P. (1991). Are 5-HT1 receptors involved in obsessive–compulsive disorder? Abstracts Spanish Biological Society of Psychiatry, Vol. 29, *5th International Congress on biological psychiatry*, Florence, 9–14 June.

Flament, M.F., Rapoport, J.L., Berg, C.J., Sceery, W., Kilts, C., Mellstram, B., Linnoila, M. (1985). Clorimipramine treatment of childhood obsessive–compulsive disorder: a double-blind controlled study. *Archives of General Psychiatry*, **42**, 977–83.

Flament, M.F., Rapoport, J.L., Murphy, D.L., Berg, C.J., and Lake, R. (1987). Biochemical changes during clomipramine treatment of childhood obsessive–compulsive disorder. *Archives of General Psychiatry*, **44**, 219–25.

Foa, E.B. (1979). Failure in treating obsessive compulsives. *Behaviour Research and Therapy*, **17**, 169–76.

Fontaine, R. and Chouinard, G. (1985). Fluoxetine in the treatment of obsessive–compulsive disorder. *Progress in Neuropsychopharmacology and Biological Psychiatry*, **9**, 605–8.

Fontaine, R. and Chouinard, G. (1986). An open clinical trial of fluoxetine in the treatment of obsessive–compulsive disorder. *Journal of Clinical Psychopharmacology*, **6**, 98–101.

Garber, H.J., Ananth, J.V., Chiu, L.C., Griswold, V.J., and Oldendorf, W.H. (1989). Nuclear magnetic resonance study of obsessive–compulsive disorder. *American Journal of Psychiatry*, **146**, 1001–5.

Gittleson, N. (1966). Depressive psychosis in the obsessional neurotic. *British Journal of Psychiatry*, **112**, 883–7.

Goodman, W.K., Price, L.H., Rasmussen, S.A., Delgado, P.L., Heninger, G.R., and Charney, D.S. (1989*a*). Efficacy of fluvoxamine in obsessive–compulsive disorder: a double-blind comparison with placebo. *Archives of General Psychiatry*, **46**, 36–44.

Goodman, W.K., Price, L.H., Rasmussen, S.A., Heninger, G.R., and Charney, D.S. (1989*b*). Serotonergic agents in obsessive–compulsive disorder, fluvoxamine as an antiobsessional agent. *Psychopharmacology Bulletin*, **25**, 31–5.

Goodman, W.K., Delgado, P.L., Price, L.H. *et al.* (1989*c*). Fluvoxamine versus desipramine in OCD. *American Psychiatry Association 142nd annual meeting*, p. 186.

Goodman, W.K., Price, L.H., Delgado, P.L., Krystal, J.H., Nagy, L.M., Rasmussen, S.A. *et al.* (1990). Specificity of serotonin reuptake inhibitors in the treatment of obsessive–compulsive disorder. Comparison of fluvoxamine and desipramine. *Archives of General Psychiatry*, **47**, 577–85.

Gordon, A. and Rasmussen, S.A. (1988). Mood related obsessive–compulsive symptoms in a patient with bipolar affective disorder. *Journal of Clinical Psychiatry*, **49**, 27–8.

Grebe, H.P. (1963). *Der grosse Duden*, Vol. 7. Etymologie Dudenverlag, Mannheim.

Greingras, M. (1977). An uncontrolled trial of clomipramine in the treatment of phobic and obsessional states in general practice. *Journal of International Medical Research* (Suppl. 5), 111–15.

Greist, J.H., Jefferson, J.W., Rosenfeld, R., Gutzmann, L.D., March, J.S., and Barklage, N.E. (1990c). Clomipramine and obsessive–compulsive disorder: a placebo controlled double-blind study of 32 patients. *Journal of Clinical Psychiatry*, **51**, 292–7.

Hamik, A. and Peroutka, S.J. (1989). 1-(m-Chlorophenyl) piperazine (mCPP) interactions with neurotransmitter receptors in the human brain. *Biological Psychiatry*, **25**, 569–75.

Hermesh, H., Hoffnung, R.A., Aizenberg, D., Molcho, A., and Munitz, H. (1989). Catatonic signs in severe obsessive–compulsive disorder. *Journal of Clinical Psychiatry*, **50**, 303–5.

Hollander, E., Schiffman, E., Cohen, B., Rivera-Stein, M.A., Rosen, W., Gorman, J.M. *et al.* (1990). Signs of central nervous systems dysfunctions in obsessive–compulsive disorder. *Archives of General Psychiatry*, **47**, 27–32.

Hudson, J.M., Pope, H.G. (1990). Affective spectrum disorder: does antidepressant respond identify a family of disorders with a common pathophysiology? *American Journal of Psychiatry*, **147**, 552–64.

Insel, T.R. (1982). Obsessive–compulsive disorder: five clinical questions and a suggested approach. *Comprehensive Psychiatry*, **23**, 241–5.

Insel, T.R. and Akiskal, H.S. (1986). Obsessive–compulsive disorder with psychotic features. A phenomenological analysis. *American Journal of Psychiatry*, **143**, 1527–33.

Insel, T.R., Kalin, N., Guttmacher, L., Cohen, R., and Murphy, D. (1982a). The dexamethasone suppression test in patients with primary obsessive–compulsive disorder. *Psychiatry Research*, **6**, 153–60.

Insel, T.R., Alterman, M.A., and Murphy, D.L. (1982b). Antiobsessional and antidepressant effects of clorimipramine in the treatment of obsessive–compulsive disorder. *Psychopharmacology Bulletin*, **18**, 115–17.

Insel, T.R., Gillin, J.C., Moore, A., Mendelson, W.B., Lowenstein, R.J., and Murphy, D.L. (1982c). The sleep of patients with obsessive–compulsive disorder. *Archives of General Psychiatry*, **39**, 1372–7.

Insel, T.R., Murphy, D.L., Cohen, R., Alterman, I., Kilts, C., and Linnoila, M. (1983). Obsessive–compulsive disorder. A double blind trial of clorimipramine and clorgyline. *Archives of General Psychiatry*, **40**, 605–12.

Insel, T.R., Mueller, E.A., Gillin, J.C., Siever, L.J., and Murphy, D.L. (1984). Biological markers in obsessive–compulsive disorders. *Journal of Psychiatric Research*, **18**, 407–23.

Insel, T.R., Mueller, E.A., Alterman, I., Linnoila, M., and Murphy, D.L. (1985). Obsessive–compulsive disorder and serotonin. Is there a connection? *Biological Psychiatry*, **20**, 1174–88.

Janet, P. (1903). *Les obsessions et la psychasthénie*. Félix Alcan, París.

Jaskari, M. (1980). Observations on mianserin in the treatment of obsessive neuroses. *Current Medical Research Opinion*, **6** (Suppl. 7), 128–31.

Jaspers, K. (1909). *Seensuch und Verbrechung*. FCW Vogel, Leipzig.

Jenike, M.A. (1981). Rapid response of severe obsessive–compulsive disorder to tranylcypromine. *American Journal of Psychiatry*, **138**, 1249–50.

Jenike, M.A. (1991*a*). Epidemiology, description and classification of obsessive–compulsive disorder. *5th World Congress of biological psychiatry*. Florence, 9–14 June.

Jenike, M.A. (1991*b*). Geriatric obsessive–compulsive disorder. *Journal of Geriatric Psychiatry and Neurology*, **4**, 34–9.

Jenike, M.A., Baer, L., Summergrad, P., Weilburg, J.B., Holland, A., and Seymour, R. (1989*a*). Obsessive–compulsive disorder: a double blind trial of clomipramine in 27 patients. *American Journal of Psychiatry*, **146**, 1328–30.

Jenike M.A., Buttolph, L., Baer, L., Ricciadi, J., and Holland, A. (1989*b*). Open trial of fluoxetine in obsessive–compulsive disorder. *American Journal of Psychiatry*, **146**, 7.

Jenike M.A., Baer, L., and Greist, J.H. (1990). Clomipramine versus fluoxetine in obsessive–compulsive disorder: a retrospective comparison of side effects and efficacy. *Journal Clinical Psychopharmacology*, **10**, 122–4.

Jimenez Garcia, P. (1967). Experiencia clínica con clorimipramina en enfermos psiquiátricos. *Libros de Becas Cursos*, **58**, 179–201.

Joffe, R.T. and Brown, P. (1984). Clinical and biological correlates of sleep deprivation in depression. *Canadian Journal of Psychiatry*, **29**, 530–6.

Joffe, R.T. and Swinson, R.P. (1988). Total sleep deprivation in patients with obsessive–compulsive disorder. *Acta Psychiatrica Scandinavica*, **77**, 483–7.

Kahn, R.S., Westenberg, H.C., and Jolles, J. (1984). Zimelidine treatment of obsessive–compulsive disorder: biological and neuropsychological aspects. *Acta Psychiatrica Scandinavica*, **69**, 259–61.

Karabanow, O. (1977). Double-blind controlled study in phobias and obsessions complicated by depression. *International Journal of Medical Research*, **5** (Suppl. 5), 42–8.

Kendell, R.E. and Discipio, W.Z. (1970). Obsessional symptoms and obsessional personality traits in patients with depressive illness. *Psychological Medicine*, **1**, 65–72.

Khanna, S. (1991). Oral *m*-CPP challenge in obsessive–compulsive disorder: behavioural variables. Biological Psychiatry 29. Abstracts from the *5th International Congress on biological psychiatry*. Florence, 9–14 June.

Kincannon, J.C. (1968). Prediction of the standard MMPI scores from 71 items. The Mini-Mult. *Journal of Consulting and Clinical Psychology*, **32**, 319–25.

Klein, D.F. (1989). *The pharmacological validation of psychiatric diagnosis, in the validity of psychiatric diagnosis* (ed. L.N. Robins and J.E. Barret. Raven Press, New York.

Kleinman, J.E., Potkin, S., Rogol, A., Buchbaum, M., Murphy D.L., and Fillin,

J.C. (1979). A correlation between platelet monoamine oxidase activity and plasma prolactin concentrations in man. *Science*, **206**, 479–81.

Kraus, A. (1977). *Sozialverhalten und Psychosen manisch-depressiver*. Enke, Stuttgart.

Kraus, A. (1980). Psychopathologie und Klinik der manisch-depressiven Psychosen. In *Die Psychologie des 20 jahrhunderts* (G. Strube), Vol. X (ed. V.H. Peters), pp. 437–64. Kindler, Zurich.

Kringlen, E. (1965). Obsession neurotics: a long-term follow up. *British Journal of Psychiatry*, **111**, 709–22.

Laakmann, G., Gugath, M., Kuss, H.J., and Zygan, K. (1984a). Comparison of growth hormone and prolactin stimulation induced by chlorimipramine and desipramine in man in connection with chlorimipramine metabolism. *Psychopharmacology (Berlin)* **82**, 62–7.

Laboucarie, J., Rascol, A., Jorda, P., Guraud, R., and Leinadier, H. (1967). New prospects in the treatment of melancholic states. Therapeutic study of a major antidepressant, chlorimipramine. *Revive Medicale de Toulouse* **3**, 863–72.

Ledesma (1977). La agresividad en la psicosis maniaco depresiva. In *Las depresiones, progresos en su diagnostico y tratamiento* (ed. Juan J. López-Ibor, Jr). Toray, Barcelona.

Legrand Du Saulle, M. (1875). *La folie du doute (avec délire du Toucher)*. Delahaye, Paris.

Leonard, H.L., Swedo, S.E., Rapoport, J.L., Koby, E.V., Lenane, M.C., Cheslow, D.L. *et al.* (1989). Treatment of obsessive–compulsive disorder with clomipramine and desipramine: a double-blind–crossover comparison. *Archives of General Psychiatry*, **46**, 1088–92.

Levine, R., Hoffman, J.S., Knepple, E.D. *et al.* (1989). Long term fluoxetine treatment of a large number of obsessive–compulsive patients. *Journal of Clinical Psychopharmacology*, **9**, 281–3.

Lewis, A. (1934). 'Melancholia: a clinical survey'. *Journal of Mental Sciences*, **80**, 227–355.

Lewis, A. (1936). Problems of obsessional illness. *Proceedings of the Royal Society of Medicine*, **36**, 325–6.

Lieberman, J. (1984). Evidence for a biological hypothesis of obsessive compulsive disorder. *Neuropsychobiology*, **11**, 14–21.

Lieberman, J.A., Kane, J.M., Sarantakos, S., Cole, K., Howard, A., Borenstein, M. *et al.* (1985). Dexamethasone suppression test in patients with obsessive–compulsive disorder. *American Journal of Psychiatry*, **142**, 747–51.

Liebowitz, M.R., Fyer, A.J., Gorman, J.M., Cambeas, R.B., Sandberg, D.P., Hollander, E. *et al.* (1988). Tricyclic therapy of the DSM-III anxiety disorders: a review with implications for further research. *Journal of Psychiatry Research*, **22**, (Suppl. 1), 7–31.

Liebowitz, M.R., Hollander, E., Fairbanks, J. *et al.* (1991). Fluoxetine for adolescents with obsessive–compulsive disorder. *American Journal of Psychiatry*, **147**, 370–1.

Lipsedge, M.S. and Prothero, W. (1987). Clonidine and clomipramine in obsessive–compulsive disorder. *American Journal of Psychiatry*, **144**, 965–6.

Lopez Ibor, J.J. (1950). *La angustia vital*. Paz Montalvo, Madrid.

Lopez Ibor, J.J. (1966). *Las neurosis como enfermedades del ánimo.* Gredos, Madrid.

Lopez-Ibor, J.J. Jr (1969). Intravenous perfusions of monochlorimipramine. Technique and results. In *The present status of psychotropic drugs* (A. Cerleti and F.J. Bové). Excerpta Medica, Amsterdam.

Lopez-Ibor, J.J. Jr (1972). *Los equivalentes depresivos.* Paz Montalvo, Madrid.

Lopez-Ibor, J.J., Jr. and López-Ibor, J.M. (1973). Tratamiento psicofarmacológico de las neurosís obsesivas. *Actas luso-españolas de Neurología, Psiquiatría y Cientas Afines*, **1**, 767–74.

Lopez-Ibor, J.J. Jr (1988). The involvement of serotonin in psychiatric disorders and behavior. *British Journal of Psychiatry*, **153** (Suppl. 3), 26–39.

Lopez-Ibor, J.J. Jr (1991*a*). The masking and unmasking of depression (ed. J.P. Feighner and W.F. Boyer). Wiley, London.

Lopez-Ibor, J.J. Jr (1991*b*). Psychopharmacological approach to obsessive–compulsive disorder. *5th International Congress on biological psychiatry*, Florence, 9–14 June.

Lopez-Ibor, J.J. Jr and Fernandez-Cordoba, F. (1967). La monoclorimipramina en enfermos resistentes a otros tratamientos. *Actas Luso-Españolas de Neurología y Psiquiatría*, **26**, 119–47.

Lopez-Ibor, J.J. Jr, Viñas Pifarre, R.; and Saiz Ruiz, J. (1980). Bases biológicas del trastorno obsesivo-compulsivo. En: Estados obsesivos, fóbicos y crisis de angustia (ed. P. Pichot, J. Giner, C. Ballu). Ediciones Aran, Madrid.

Lopez-Ibor, J.J. Jr, Saiz-Ruiz, J., and Moral, L. (1988). The fenfluramine challenge test as an index of severity of the affective disorders. *Pharmacopsychiatry*, **21**, 9–14.

Luxemberg, J.S., Swedo, S.E., Flament, M.F., Fredland, R.P., Rapoport, J., and Rapoport, S.I. (1988). Neuroanatomical abnormalities in obsessive–compulsive disorder detected with quantitative X-ray computed tomography. *American Journal of Psychiatry*, **145**, 1089–93.

Malloy, P., Rasmussen, S., Braden, W., and Haier, R.J. (1989). Topographic evoked potential mapping in obsessive–compulsive disorder: evidence of frontal lobe dysfunction. *Psychiatry Research*, **28**, 63–71.

Marks, I.M. (1981). Review of behavioral psychotherapy in obsessive–compulsive disorder. *American Journal of Psychiatry*, **138**, 584–92.

Marks, I.M. (1983). Are there anticompulsive or antiphobic drugs? Review of the evidence. *British Journal of Psychiatry*, **143**, 338–47.

Marks, I.M. (1987). Obsessive–compulsive disorder. In *Fears, phobias and rituals: panic anxiety and their disorders*, pp. 423–56. Oxford University Press.

Marks, I.M., Hodson, R., and Rachman, S. (1975). Treatment of chronic obsessive–compulsive disorder two years after *in vivo* exposure. *British Journal of Psychiatry*, **127**, 349–67.

Marks, I.M., Stern, R.S., Manson, D., Cobb, J., and McDonald, R. (1980). Clomipramine and exposure for obsessive–compulsive rituals. *American Journal of Psychiatry*, **136**, 1–25.

Marshall W.K. (1971). Treatment of obsessional illness and phobic anxiety state with clomipramine. *British Journal of Psychiatry*, **119**, 467–8.

Mavissakalian, M. and Michelson, L. (1983). Tricyclic antidepressants in obsessive–compulsive disorder. Antiobsessional or antidepressant agents? *Journal of Nervous and Mental Disease*, **171**, 301–6.

Mavissakalian, M., Turner, S.M., Michelson, L., and Jacob, R. (1985). Tricyclic antidepressants in obsessive–compulsive disorder. Antiobsessional or antidepressant agents? II. *American Journal of Psychiatry*, **142**, 572–6.

Mavissakalian, M., Jones, B., and Olson, S. (1990*a*). Absence of placebo response in obsessive–compulsive disorder. *Journal of Nervous and Mental Disease*, **178**, 268–70.

Mavissakalian, M., Jones, B., Olson, S., and Perel, J.M. (1990*b*). Clomipramine in obsessive–compulsive disorder: clinical response and plasma levels. *Journal of Clinical Psychopharmacology*, **10**, 261–8.

Monteiro, W., Marks, I.M., Noshirvani, H., and Checkleys, S. (1986). Normal dexamethasone suppression test in obsessive–compulsive disorder. *British Journal of Psychiatry*, **134**, 382–9.

Montgomery, S.A. (1980). Clomipramine in obsessional neurosis: a placebo-controlled trial. *Pharmaceutical Medicine*, **1**, 189–92.

Montgomery, S.A. and Asberg, M. (1979). A new Depression Scale designed to be sensitive to change. *Journal of Psychiatry*, **148**, 326–9.

Montgomery, S.A. and Fineberg, N. (1987). New findings in obsessive–compulsive disorder. Autumn Quarterly Meeting, The Royal College of Psychiatrists.

Montgomery, S.A., Fineberg, N., and Montgomery, D.B. (1990). The efficacy of serotonergic drugs in obsessive–compulsive disorder—power calculations compared with placebo. In *Current approach: obsessive–compulsive disorder*. Ashford Colour Press.

Mueller, E.A., Murphy, D.L., and Sunderland, T. (1985). Neuroendocrine effects of *m*-chlorophenylpiperazine, a serotonin agonist, in humans. *Journal of Clinical Endocrinology and Metabolism*, **61**, 1179–84.

Murphy, D.L. (1988). The newer serotonergic drugs in anxiety, depression, obsessive–compulsive disorder, and related psychiatric disorders. Presented at the *2nd Congress of the Spanish biological psychiatry*, Madrid, November.

Murphy, D.L., Siever, L.J., and Insel, T.R. (1985). Therapeutic responses to tricyclic antidepressants and related drugs in non-affective disorder patients population. *Progress in Neuro-Psychopharmacology and Biological Psychiatry*, **9**, 3–13.

Noshirvani, H.F., Kasvikis, Y., Marks, I.M., Tsakiris, F., and Monteiro, W.O. (1991). Gender-divergent etiological factors in obsessive–compulsive disorder. *British Journal of Psychiatry*, **158**, 260–3.

Oreland, L., Niberg, A., Aseberg, M., Traskman, L., Sjostrand, L., Thoren, P. *et al.* (1981). Platelet activity and monoamine metabolites in cerebrospinal fluid in depressed and suicidal patients and in healthy controls. *Psychiatry Research*, **4**, 21–4.

Pato, M.T., Pigott, T.A., Hill, J.L. Grover, G.N., Bernstein, S., and Murphy, D.L. (1991). Controlled comparison of buspirone and clomipramine in obsessive–compulsive disorder. *American Journal of Psychiatry*, **148**, 127–9.

Perse, T.L., Greist, J.H., Jefferson, J.W., Rosenfeld, R., and Dar, R. (1987). Fluvoxamine treatment of obsessive–compulsive disorder. *American Journal of Psychiatry*, **144**, 1543–8.

Pigott, T.A., Pato, M.T., Bernstein, S.E., Grover, G.N., Hill, J.L., Tolliver, T.J. *et al.* (1990). Controlled comparisons of clomipramine and fluoxetine in the treatment of obsessive–compulsive disorder. *Archives of General Psychiatry*, **47**, 926–32.

Prasad, A. (1984). A double blind study of Imipramine versus Zimelidine in treatment of obsessive–compulsive neurosis. *Pharmacopsychiatry*, **17**, 61–2.

Prasad, A. (1985). Efficacy of trazodone as an antiobsessional agent. *Pharmacology and Biochemical Behaviour*, **22**, 347–8.

Price, L.H., Goodman, W.K., Charney, D.S., Rasmussen, S.A., and Heninger, G.R. (1987). Treatment of severe obsessive–compulsive disorder with fluvox-amine. *American Journal of Psychiatry*, **144**, 1059–61.

Pulman, M.D., Yassa, R., and Anath, J. (1984). Clomipramine treatment of repetitive behavior. *Canadian Journal of Psychiatry*, **29**, 254–5.

Rapoport, J. and Wise, S.P. (1988). Obsessive–compulsive disorder: evidence for basal ganglia dysfunction. *Psychopharmacology Bulletin*, **24**, 380–4.

Rapoport, J., Elkins, R., *et al.* (1980). Clinical controlled trial of clomipramine in adolescents with obsessive–compulsive disorder. *Psychopharmacology Bulletin*, **3**, 61–3.

Rapoport, J., Elkins, R., Langer, D.H., Sceery, W., Buchsbaum, M.S., Gillin, J.C. *et al.* (1981). Childhood obsessive–compulsive disorder. *American Journal of Psychiatry*, **138**, 1545–54.

Rasmussen, S.A. (1984). Lithium and tryptophan augmentation in clomipramine resistant obsessive–compulsive disorder. *American Journal of Psychiatry*, **141**, 1283–5.

Rasmussen, S.A. and Eisen, J.L. (1988). Clinical features and phenomenology of obsessive–compulsive disorder. *Psychiatric Annals*, **19**, 67–73.

Rasmussen, S.A. and Eisen, J.L. (1990). Epidemiology of obsessive–compulsive disorder. *Journal of Clinical Psychiatry*, **51** (Suppl.), 10–14.

Rasmussen, S.A. and Tsuang, M.T. (1986a). Epidemiology and clinical features of obsessive–compulsive disorder. In *Obsessive–compulsive disorders, theory and management* (ed. M.A. Jenike, L. Baer, and W.E. Minichiello). PSG, Littleton MA.

Rasmussen, S.A. and Tsuang, M.T. (1986b). Clinical characteristics and family history in DSM-III obsessive–compulsive disorder. *American Journal of Psychiatry*, **143**, 317–22.

Renynghe De Voxrie, G.U. (1968). L'Ananfranil dans l'obsessions. *Acta Neuro-logica Belgica*, **68**, 787–92.

Riddle, M.A., Hardin, M.T., King, R., Scahill, L. and Woolston, J.L. (1990). Fluoxetine treatment of children and adolescents with Torette's and obsessive compulsive disorders: preliminary clinical experience. *Journal of the Academy of Child and Adolescent Psychiatry*, **29**, 45–8.

Rosenberg, C.M. (1968). Complications of obsessional neurosis. *British Journal of Psychiatry*, **114**, 177–478.

Roth, M., Gurney, C., and Garside, R.F. (1972). Studies in the clasificatios of affective disorders. The relationship between anxiety states and depressive illness. *British Journal of Psychiatry*, **121**, 147–61.

Roy-Byrne, P.P., Uhde, T.W., Post, R.M. and Joffe, R.T. (1984). Relationship of response to sleep deprivation and carbamazepine in depressed patient. *Acta Psychiatrica Scandinavica*, **69**, 379–82.

Schneider, K. (1950). *Klinische psychopathologie* (3rd edn). Thieme, Stuttgart.

Siever, L.J., Insel, J.R., Jimerson, D.C., Lake, C.R., Uhde, T.W., Aloi, J. *et al.* (1983). Growth hormone response to clonidine in obsessive–compulsive patients. *British Journal of Psychiatry*, **142**, 184–6.

Solyom, L. and Sookman, D. (1977). A comparison of clomipramine hydrochloride and behavioural therapy in the treatment of obsessive neurosis. *Journal of International Medical Research*, **5** (Suppl. 5), 49–61.

Stern, R.S., Marks, I.M., Wright, J., and Luscombe, D.K. (1980). Clomipramine: plasma levels, side effects and outcome in obsessive–compulsive neurosis. *Postgraduate Medical Journal*, **56** (Suppl. 1), 134–9.

Swedo, S.E., Schapiro, M.B., Grady, Ch. L., Chelson, D.L., Leonard, H.L., Kumar, A. *et al.* (1989). Cerebral glucose metabolism in childhood-onset obsessive–compulsive disorder. *Archives of General Psychiatry*, **46**, 518–23.

Swinson, R.P. and Joffe, R.T. (1988). Biological challenges in obsessive–compulsive disorder. *Progress in Neuro-Psychopharmacology and Biological Psychiatry*, **12**, 269–75.

Thoren, P., Äsberg, M., Cronholm, B., Jörnestedt, L., and Träskman, L. (1980*a*). Clomipramine treatment of obsessive–compulsive disorder I: a controlled trial. *Archives of General Psychiatry*, **37**, 1281–8.

Thoren, P., Äsberg, M., Bertilsson, L., Mellström, B., and Sjöqvist, F. (1980*b*). Clomipramine treatment of obsessive–compulsive disorder II: Biochemical aspects. *Archives of General Psychiatry*, **37**, 1289–94.

Towey, J., Bruder, G., Hollander, E., Friedman, D., Erham, H., Liebowitz, M. *et al.* (1990). Endogenous event related potentials in obsessive–compulsive disorder. *Biological Psychiatry*, **28**, 92–8.

Turner, S.M., Jacob, R.G., Beidel, D.C. *et al.* (1985). Fluoxetine treatment of obsessive–compulsive disorder. *Journal of Clinical Psychopharmacology*, **5**, 207–12.

Vallejo, J., Olivares, J., Marcos, T., Martinez-Osaba, M.J., Ribera, F., and Bulbena, A. (1988). Dexamethasone suppression test and primary obsessional compulsive disorder. *Comprehensive Psychiatry*, **29**, 498–502.

Vallejo, J., Olivares, J., Marcos, T., Bulbena, A., and Otero, A. (1989). *Clomipramine vs phenelzine: double-blind trial in obsessive disorders*. Excerpta Medica. International Congress Series 899. VIIIth International Congress of Psychiatry, Athens, 12–19 October.

Vaughan, M. (1976). The relationship between obsessional personality, obsessions in depression and symptoms of depression. *British Journal of Psychiatry*, **129**, 36–9.

Videbach, T.H. (1975). The psychopathology of anancastic endogenous depression. *Acta Psychiatrica Scandinavica*, **52**, 336–73.

Viñas, R. (1991). Trastorno obsesivo–compulsivo: estudio clínico y neuroendicrino. Doctoral thesis. Universidad Autónoma de Madrid.

Volankas, J., Nezirogli, F., and Yaryura-Tobias, J.A. (1985). Clomipramine and imipramine in obsessive–compulsive disorder. *Psychiatry Research*, **14**, 83–91.

Warneke, L.B. (1985). Intravenous chlorimipramine in the treatment of obsessional disorder in adolescence, case report. *Journal of Clinical Psychiatry*, **46**, 100–3.

Waxman, D. (1973). A general practice investigation into the use of anafranil in phobic and obsessional disorders. *Journal of International Medical Research*, **1** (Suppl. 5), 417–20.

Waxman, D. (1975). An investigation into the use of Anafranil in phobic and obsessional disorders. *Scottish Medical Journal*, **20**, 61–6.

Waxman, D. (1977). A clinical trial of clomipramine and diazepan in the treatment of phobic and obsessional illness. *Journal of International Research* (Suppl. 5), 99–110.

Welner, A., Reich, T., and Robins, E. (1976). Obsessive–compulsive neurosis: record, followup, and family studies. I. Inpatient record study. *Comprehensive Psychiatry*, **17**, 527–39.

WHO (World Health Organization) (1978). *Mental disorders (ICD-9).* Glossary and guide to their classification. WHO, Geneva.

WHO (World Health Organization) (1987). *Mental Disorders. Draft proposal for 10th edn. (ICD-10).* WHO, Geneva.

Wyndowe, J., Solyom, L., and Anath, J. (1975). Anafranil in obsessive–compulsive neurosis. *Current Therapeutic Research*, **18**, 611–17.

Yaryura-Tobias, J.A. and Bhagvan, H.N. (1977). L-trytophan in obsessive–compulsive disorder. *American Journal of Psychiatry*, **234**, 1298–9.

Yaryura-Tobias, J.A., Neziroglu, F., and Bergman, L. (1976). Clomipramine for obsessive neurosis: and organic approach. *Current Therapeutic Research*, **20**, 541–7.

Zohar, J. and Insel, T.R. (1987a). Drug treatment of obsessive–compulsive disorder. *Journal of Affective Disorders*, **13**, 193–202.

Zohar, J. and Insel, T.R. (1987b). Obsessive–compulsive disorder: psychobiological approaches to diagnosis, treatment and pathophysiology. *Biological Psychiatry*, **22**, 667–87.

Zohar, J., Mueller, E.A., Insel, T.R., Zohar-Kadouch, R.C., and Murphy, D.L. (1987c), Serotonergic responsivity in obsessive–compulsive disorder: comparison of patients and healthy controls. *Archives of General Psychiatry*, **44**, 946–51.

13

The place of antidepressants in long-term treatment

E.S. PAYKEL

LONGER TERM OUTCOME IN DEPRESSION

As a result of more than 30 years of research since the development of modern antidepressant drugs, the acute treatment of depression can now reasonably be regarded as well worked out. The places of ECT; tricyclics, MAO inhibitors, other antidepressant drugs, and dynamic psychotherapies, are now relatively clear (Paykel 1989), although the precise indications for the relative newcomer, cognitive therapy, still remain to be clarified.

The outcome of acute treatment is relatively good. Most patients improve considerably within a few months of presentation, although in a proportion remission is delayed or there are residual symptoms and a small number of patients remain severely ill and treatment-resistant. However it has also become apparent that the longer term outcome is far less satisfactory. I will briefly deal with this problem, before considering the role of antidepressants in its prevention.

First we need to consider the distinction between early and late symptom return. Frank *et al.* (1991) have pointed out a number of key terms in a sequence of the course of treatment in depression: (1) *Remission*—the earlier part of the phase when symptoms have gone, which may be interrupted by (2) *Relapse*—early return of symptoms. Hopefully this may be replaced or followed by (3) *Recovery*—restitution to normality, with resolution of the underlying disorder, which may, unfortunately, be followed by (4) *Recurrence*—the occurrence of a new episode.

The distinction which has been widely accepted in the literature is that between early relapse and later recurrence. It has become important since evidence has accumulated that early withdrawal of antidepressants after remission of symptoms is associated with high rates of symptom return. This can plausibly be seen as the return of the original episode in which the symptoms have been suppressed by treatment, but the underlying biological process may not have recovered. It has been assumed that given time this recovery does happen, either due to treatment or spontaneously and that later symptom return represents occurrence of a new episode.

Although this distinction is plausible and heuristically useful in guiding the planning and interpretation of research and the construction of treatment guidelines, it should be noted that it has not been fully validated, other than by the observation that the frequency of symptom return declines after the first year of follow up (Lavori *et al*. 1984). Nor is there any firm rule as to when a relapse becomes a recurrence, although most investigators would consider something like 6 months after response as the cut-off point.

It is also important to note that the relapse/recurrence distinction is not the same as another one with which it has sometimes been confused: that between those studies starting treatment during a maintenance phase and those adopting discontinuation designs, involving comparisons between symptom return in patients continued on pre-existing treatment with antidepressant or lithium, and those withdrawn from it. Most studies of relapse have been discontinuation studies but not all, for example a study in which imipramine or placebo was given from the start with ECT (Seager and Bird 1962). Studies of recurrence often involve discontinuation also. It is even possible for an early discontinuation trial to study both relapse and recurrence, if the subsequent follow-up period is long. This becomes important in interpreting studies.

FOLLOW-UP STUDIES

There have been many short-term and long-term follow-up studies. Coryell and Winokur (1992) have recently reviewed them comprehensively. Bipolar disorder is more recurrent than unipolar.

I will concentrate on unipolar depression, since it is the more common problem, and more relevant to long-term use of antidepressants. What one would really like to know is the outcome of depression in representative samples treated since the advent of the full range of modern treatments, i.e. since the 1960s. These are not fully available: even in recent studies, the samples are somewhat selective, and tend to focus on the more severely ill, those treated as inpatients, and patients referred to special University treatment centres who may have been resistant to earlier treatment.

The most informative short-term follow up is that conducted by Keller and colleagues from the NIMH collaborative study of the Psychobiology of Depression (Keller *et al*. 1982*a,b*, 1983). The sample was predominantly inpatient, with about one-quarter outpatients, and predominantly from special centres. Although episode length was protracted if taken from onset, the rate of remission from the time of treatment was moderately good: 63 per cent of patients recovered within the first 4 months, after which recovery rates declined markedly and 26 per cent of patients still had not recovered at 1 year. The recovery criterion was fairly stringent, and given this, and the nature of the sample, this was not a bad outcome. Outcome was worse where the episode was superimposed on what would now be regarded as dysthymic disorder, where onset was slower and where severity was greater.

When patients were followed beyond the point of remission, outcome was not so good. 12 per cent of patients relapsed within the month after recovery, and 24 per cent within the first three months. 36 per cent did so in the first year. Relapse was associated with more previous episodes, a dysthymic pattern, and older age of onset. These figures are consistent with those which emerged from the studies of drug continuation versus withdrawal undertaken in the 1970s, which will be reviewed in a later section.

Longer term follow-up studies show substantial recurrence rates. Until recently however, most samples receiving long-term follow up had been treated in a previous era, and it was difficult to know if findings applied in the circumstances of modern acute and maintenance treatment. This is no longer the case. At the end of 1988 two long-term follow-up studies were published, one of 89 depressives admitted to the Maudsley Hospital in 1965–6, originally studied by R.E. Kendell and followed up 18 years later (Lee and Murray 1988); the other of 145 admissions to an Australian University Hospital between 1966–70 and followed up 15 years later (Kiloh et al. 1988). Outcomes were poor. Over approximately 16 years, 60 per cent and 56 per cent respectively were readmitted at least once: this applied to 51 per cent and 47 per cent of those who were first admissions at index episode. Readmission was more common in endogenous than in neurotic depressives. Among those readmitted, some showed very recurrent illnesses. Among those not readmitted 22 per cent and 16 per cent showed either mild/moderate chronic illness, or very slow recovery from index episode. Overall, in the two studies, respectively 25 per cent and 11 per cent showed very poor outcomes with chronic severe disorder and handicaps. Suicide and overall mortality rates were increased compared with the general population. An 11-year follow-up of a Canadian sample produced comparably poor outcome (Lehmann et al. 1988). The classic review of suicide in affective disorders by Guze and Robins (1970) suggested that 15 per cent of deaths are by suicide, and there is good evidence that overall mortality rates are increased (Coryell and Winokur 1992).

What one would also like to know is the long-term outcomes in outpatients, general practice patients, and subjects in the community. These cannot be as bad, or the episode rate and lifetime incidence figures from the community would suggest that everybody is depressed for much of the time. What is noteworthy from the two inpatient follow ups is that, among first episode patients, half had at least one further episode requiring admission, i.e. of comparable intensity to the first episode, and half did not; a smaller number had very recurrent episodes; some additional subjects had at least one further milder recurrence not requiring admission. A clinical guess would be that whatever the severity and intensity of care: community, general practice, outpatient, inpatient, about 50 per cent have at least one further episode at that level and 50 per cent do not, although they may have another milder episode. This is not very well informed guesswork, however; it needs testing properly.

An additional problem revealed in the above is chronicity of symptoms. This may be a particular problem in the outpatient range. The follow-up studies suggest that many patients treated initially as inpatients or outpatients improve only partially, with residual symptoms which continue in the community. Evidence suggests the existence of moderate numbers of patients with chronic and fluctuating pictures in outpatient clinics and in general practice (Paykel and Griffith 1983).

In keeping with this has been the diagnostic recognition of chronic depressive syndromes, especially that of dysthymia in DSM-III and DSM-III-R. This syndrome of mild fluctuating depression, with spells of remission which are not long lasting, and a chronic or lifelong pattern, appears more closely related to major depression than has hitherto been realized. A high proportion of dysthymics ultimately do develop a major depression (ACNP 1990), a pattern that has sometimes been described as 'double depression' (Keller and Shapiro 1982). DSM-III-R, unlike its predecessors, excluded from the diagnosis of dysthymia those cases where the symptoms immediately followed major depression. This appears in retrospect artificial, and dysthymia overlaps with the chronic residual symptoms previously described for some depressives.

STUDIES OF CONTINUATION THERAPY IN RELAPSE PREVENTION

There have been a number of controlled trials designed to test reduction in relapse rates by antidepressants. Findings are summarized in Table 13.1. The key set of studies was published mainly in the 1970s and examined, in depressives responding to initial treatment, the effects on early relapse of continuation of tricyclic antidepressant for approximately 6 months, as opposed to its early withdrawal (Mindham *et al.* 1973; Prien *et al.* 1973*a*, 1984; Klerman *et al.* 1974; Paykel *et al.* 1975; Coppen *et al.* 1978*a*; Stein *et al.* 1980). The study by Prien *et al.* (1973*a*) involved both relapse and recurrence, but separated the findings. Findings shown are for unipolar disorder. In bipolar disorder only lithium was effective. Prien *et al.* (1984) also studied recurrence and bipolars but results shown are for unipolars only and for a separate analysis of the first 8 weeks after withdrawal (Prien and Kupfer 1986). All the studies showed benefit from continuation: relapse

TABLE 13.1. Studies of antidepressants in prevention of relapse

Study	Drug	Length of drug treatment prior to withdrawal (approximate)	Length of follow up after withdrawal	Relapse rate (%)	
				Placebo	Drug
Discontinuation designs					
Tricyclics					
Mindham et al. 1973	Amitriptyline/imipramine	3–10 weeks	6 months	50	22
Klerman et al. 1974 Paykel et al. 1975	Amitriptyline	3 months	6 months	29	12
Coppen et al. 1978a	Amitriptyline	Not described	1 year	31	0
Stein et al. 1980	Amitriptyline	6 weeks	6 months	69	28
Prien et al. 1973a*	Imipramine, lithium	Not described	4 months	73	32 / 30
Prien et al. 1984	Amitriptyline, lithium	Variable	2 months	38	5
Prien and Kupfer 1986*	imipramime and lithium				38 / 11
MAO inhibitors					
Davidson and Raft 1984	Phenelzine	2 months	5 months	100	14
Harrison et al. 1986	Phenelzine	Approx. 3 months	6 months	100	20
Nondiscontinuation designs, with ECT					
Seager and Bird 1962	Imipramine		6 months	60	17
Imlah et al. 1965	Imipramine, phenelzine		6 months	51	12

* Acute treatments varied: followed by stabilization on maintenance drugs then withdrawal double blind.

rates on discontinuation ranged from 29 to 73 per cent and were halved or more by continuing drug, although not usually completely abolished, and remaining near 30 per cent in two studies (Mindham *et al.* 1973; Stein *et al.* 1980). The varying relapse rates presumably reflected initial characteristics of the samples. The Coppen study lasted 12 months but relapses were predominantly in the first 6 months. Differences only reached significance when relapsers in amitriptyline noncompliers were excluded.

There have been fewer studies of MAO inhibitors or of newer anti-depressants, although some studies of recurrence prevention using the latter drugs also involve earlier relapse, and will be described in due course. Similar findings to the above have been obtained more recently for phenelzine (Davidson and Raft 1984; Harrison *et al.* 1986). There are hints that worsening may be more likely when MAOIs are stopped than for tricyclics, especially after long-term therapy (Tyrer 1984). If this is true, it is not clear whether it represents an effect of the drug, of the type of patient responding selectively to MAO inhibitors, or of the kind of chronic and recurrent disorders which are more likely to be treated with MAO inhibitors, because of their role as second choice drugs. In both the small continuation trials 100 per cent of patients withdrawn from MAOI relapsed, suggesting that this does have something to do with the effect of the drug.

Not all relapse prevention studies have adopted discontinuation designs. Seager and Bird (1962) treated patients receiving ECT with in addition either imipramine or placebo, and continued treatment for 6 months after discharge. There was no effect on the number of ECT treatments needed or the response rate, but the relapse rate over 6 months was considerably higher in the placebo group. Imlah *et al.* (1965) obtained similar findings in a three-group study using imipramine, phenelzine, or placebo.

Our own Yale–Boston study (Paykel *et al.* 1976) was the first of the continuation versus withdrawal antidepressant studies undertaken, although its scale and the addition of psychotherapy meant that it was not the first completed. Depressed women were treated with amitriptyline acutely and 150 responders were randomized to continue antidepressant for 8 months or withdrawn from it after 2 months either double blind or openly, in a 3 × 2 factorial design in which half received individual psychotherapy from social workers. We found relapse rates of 29 per cent on withdrawal to placebo and 27 per cent to no medication, reduced to 12 per cent by antidepressant continuation (Klerman *et al.* 1974). Residual symptoms were also reduced (Paykel *et al.* 1975). Weight gain and carbohydrate craving emerged as undesirable long-term side effects (Paykel *et al.* 1973). Psychotherapy had no effect in relapse or symptom ratings, but did improve social adjustment and interpersonal relationships. Inter-current life events precipitated relapse, but the effect of the antidepressant was not particularly in these circumstances (Paykel and Tanner 1976).

This, and the absence of psychotherapy effect, argue that the relapses were particularly associated with early drug withdrawal, a point to which I will return in due course.

This phase of short-term maintenance treatment after the acute phase has now come to be known as continuation therapy (Mindham *et al.* 1973; Prien 1992), and should be routine after antidepressant response. Its optimal length is uncertain, although 4 months after the patient is symptom free or back to the usual level of interepisode function has been suggested, and adopted in recent studies aimed at distinguishing recurrence from relapse (Frank *et al.* 1990; Robinson *et al.* 1991; Prien 1992). In a large NIMH collaborative study (Prien and Kupfer 1986) and in a smaller study (Georgotas and McCue 1989) withdrawal was only safe under these circumstances and in a longer term study by Frank *et al.* (1990) withdrawal at this point produced a low rate of worsening in the subsequent 2 months. In one of the early drug continuation trials, benefit of drug continuation was seen particularly in those patients with residual symptoms (Mindham *et al.* 1973), and this may provide an indication for longer continuation until these symptoms have subsided.

MAINTENANCE DRUG TREATMENT AND RECURRENCE

Table 13.2 sets out comparisons of antidepressants and placebo in prevention of recurrence. Many of the studies used previously recurrent samples so that subsequent recurrence rates were high. Antidepressants found superior to placebo include imipramine, maprotiline, phenelzine, zimeldine, fluoxetine, sertraline, while the one small study using nortriptyline did not find it superior to placebo. Two studies used very small samples: one (Glen *et al.* 1984) only reported significance for amitriptyline and lithium combined; in the other (Kane *et al.* 1982) imipramine was no better than placebo. It is noteworthy that two of the newer drugs, zimeldine and nomifensine, have since been withdrawn from the market because of immunological side effects not apparent in first use, which should provide a warning against the use of newer drugs rather than older and well tried ones in the situation of long-term maintenance, where exposure to adverse effects will be higher. One issue of concern is that most of the studies of newer drugs were reported in unrefereed journals or supplements. For one additional study of nomifensine quoted in a number of reviews references are to a conference abstract or to a paper not obtainable in this country and I have not included it in the table.

Almost all the studies involved discontinuation designs in subjects who had responded to initial treatment with the drug, the exceptions being Bjork (1983) who studied recurrent patients who were either on lithium

TABLE 13.2. *Studies of antidepressants in prevention of recurrence*

Study	Drug	Length of drug treatment prior to withdrawal (approximate)	Length of follow up after withdrawal	Recurrence rate (%) Placebo	Recurrence rate (%) Drug
Tricyclics					
Prien et al. 1973a	Imipramime	Not described	2 years	85	29*
	lithium				41
Prien et al. 1984	Imipramine	> 2 months	2 years	71	41†
	lithium	symptom free			57
	combination				31
Kane et al. 1982‡	Imipramine	6 months	1 year	83	100
	lithium		approx.		29
	combination				13
Glen et al. 1984	Amitriptyline	Not described	3 years	89	50
	lithium				42
Rouillon et al. 1989	Maprotiline 75 mg	2 months	1 year	35	16
	Maprotiline 37.5 mg				24
Georgotas et al. 1989	Nortriptyline	6 months	1 year	63	54
	phenelzine				13
Frank et al. 1990	Imipramine	6 months	3 years	78	22
	interpersonal				64
	psychotherapy combination				24
MAO inhibitors					
Georgotas et al. 1989	Nortriptyline (see above)				
Robinson et al. 1991	Phenelzine 60 mg	6 months	2 years	81	26
	Phenelzine 45 mg				33
Newer drugs					
Bjork 1983	Zimeldine	Euthymic for 4 months	18 months	84	32
Montgomery 1988	Fluoxetine	6 months	1 year	57	26
Doogan and Caillard (1992)	Sertraline	2 months	10 months	41	8

* Months 5–24.
† See text.
‡ Small samples.

(thus biasing against antidepressant) or had had several episodes but had not received antidepressants in the last month, and the two studies by Prien and colleagues, in which initial treatment was diverse but a partial discontinuation design was employed in that variously treated patients were stabilized on the active study drugs before withdrawal. A number of the studies used short or undefined acute treatment periods so that issues of relapse and recurrence were fused (Prien et al. 1973a, 1984; Glen et al. 1984; Rouillon et al. 1989; Doogan and Caillard, in press). This raises problems of interpretation even where data are presented separately for early relapse and later recurrence, since patients relapsing early are not at risk subsequently. Prien et al. (1973a) presented data reporting for the first 4 months and the remainder of the 2 years, and imipramine had significant effects in both periods. The data from Prien et al. (1984) for the first 8 weeks after withdrawal were analysed and presented in a subsequent paper (Prien and Kupfer 1986); more than half of the symptom returns on placebo occurred during this time (see Table 13.1). By subtraction it would appear that recurrence rates thereafter were placebo (33 per cent), imipramine (36 per cent), lithium (19 per cent), combination (20 per cent) suggesting that only lithium had an effect. The data presented by Rouillon et al. (1989) suggest symptom returns spread out equally over the year of the follow up. Among the studies postponing withdrawal until after the continuation period, only Robinson et al. (1991) found a marked tendency for relapses to occur early after withdrawal (the first 8 weeks); the other studies, where timing of recurrences is reported, suggest recurrences evenly spread out over about a year, with relatively few thereafter.

LITHIUM IN MAINTENANCE TREATMENT OF UNIPOLAR DISORDERS

There is evidence that lithium is also effective in the long-term prevention of unipolar recurrent disorders, and this must be taken into account in drawing practical therapeutic conclusions. Six studies have shown lithium superior to placebo in maintenance treatment of unipolars (Baastrup et al. 1970; Coppen et al. 1971; Prien et al. 1973a, 1984; Fieve et al. 1976; Kane et al. 1982). In a seventh study Glen et al. (1984) in a small sample found lithium and amitriptyline in a pooled analysis, better than placebo. Coppen et al. (1981) also found it superior to placebo after ECT in a 1-year study which was intermediate between continuation and maintenance. Medication was started during ECT and went on for a year. Lithium showed benefit in preventing affective morbidity: the effect was more marked in the second 6 months of the study, and was only significant during this phase. Not all these studies had adequate periods of remission to disguish continuation/relapse from prophylaxis/recurrence. In several of the studies subjects may have already had long periods on lithium and it is likely therefore that good lithium responders that were being discontinued.

The choice of long-term treatment between tricyclics and lithium in unipolar depression is not clear from the above. In direct comparisons lithium has been found superior to antidepressant in three studies (Coppen *et al.* 1976, 1978*b*; Kane *et al.* 1982) (the two Coppen studies involved discontinuation in lithium-treated samples and were probably therefore biased in its favour); equal in two (one with a small sample) (Prien *et al.* 1973*a*; Glen *et al.* 1984), inferior in one (Prien *et al.* 1984). These studies all involved imipramine or amitriptyline, except for Coppen *et al.* (1978*b*) in which the antidepressant was mianserin, and Coppen *et al.* (1976) in which it was maprotiline. Three studies have involved lithium-tricyclic combinations: in the two shown in Table 13.2 (Kane *et al.* 1982; Prien *et al.* 1984) the combination appeared more effective than either drug alone, although not necessarily significantly so. In a third study (Johnstone *et al.* 1990) the addition of lithium to amitriptyline made no difference.

This might seem still to leave the choice doubtful, and render a little illogical the undoubted fact that moderately large numbers of patients are maintained on antidepressants. In reality I do not think that it is. The common situation in which maintenance therapy is initiated in unipolar depression is immediately after acute treatment for an episode. The patient is very likely to have received an antidepressant and responded to it, rendering continuation the obvious choice, as it is in the other common situation, where symptoms return during or soon after withdrawal of the antidepressant following a continuation phase. It is only after ECT or failure of antidepressant maintenance that lithium is likely to be initiated, and even here patients are in my experience more resistant to starting it and its accompanying blood tests, than to receiving an antidepressant, with which they will already be familiar.

In the above studies, as in the continuation studies aimed at relapse prevention, although superiority to placebo was demonstrated, efficacy was incomplete. Recurrence rates on imipramine in studies showing it superior to placebo ranged from 21 per cent (Frank *et al.* 1990) to 48 per cent (Prien *et al.* 1973*a*); on lithium in studies superior to placebo they ranged from 27 per cent (Kane *et al.* 1982) to 48 per cent (Prien *et al.* 1973*b*). Patient compliance may also be a problem (Jacob *et al.* 1984).

BIPOLAR DISORDER

The place of lithium in bipolar disorder is outside the scope of this paper. There is good evidence of its superiority to placebo in preventing both manic and depressive episodes (Baastrup *et al.* 1970; Coppen *et al.* 1971; Cundall *et al.* 1972; Prien *et al.* 1973*a,b*, 1984; Fieve *et al.* 1976; Kane *et al* 1982). It is clearly the first choice treatment, with a secondary place for carbamazepine and valproate (Prien 1992).

What of antidepressant? There have only been a small number of controlled trials: they suggest that tricyclics are not of benefit alone, and precipitate mania, although the situation is equivocal for lithium-antidepressant combinations. Prien *et al.* (1973*a*) found more manic attacks on imipramine than placebo (or on lithium), although there was benefit comparable to that of lithium for depressive attacks. Kane *et al.* (1982) found imipramine no better than placebo in bipolar II patients. Three studies (Prien *et al.* 1973*a*, 1984; Kane *et al.* 1982) all found imipramine inferior to lithium. In comparisons involving combination, Kane *et al.* (1982) found combination superior to imipramine and comparable to lithium, while Prien *et al.* (1984) found combination equivalent overall to lithium, but a little superior to either drug where the index episode had been depressive (Shapiro *et al.* 1989). Again there was a high incidence of mania on imipramine alone.

The trials therefore do not indicate a place for antidepressants in maintenance treatment of bipolars. Most clinicians however still do have some patients who do not respond to other treatments and seem to derive at least limited benefit from tricyclic-neuroleptic combinations.

INDICATIONS FOR MAINTENANCE

It is not the place of this paper to attempt clinical guidelines. Angst (1981) suggested that for lithium prophylaxis of unipolar disorder the indication was one episode or more in addition to the present one, in the last 5 years. The recommendation was derived from a study of the course of the disorder in which this was found a strong predictor of two more episodes in the next 5 years. Coppen and Abou Saleh (1983) endorsed this view. It does not tell us how long to continue, but if broadened into an indication for continuing antidepressant for a year rather than 6 months after the attack, rather than necessarily an indication for lithium maintenance, I would find it reasonable. The NIMH/NIH Consensus Development Conference Statement on Pharmacologic Prevention of Recurrence (1985) did not take a definite stand, but listed a number of factors to be considered, as well as disadvantages.

Among the disadvantages are the side effects. Tricyclics tend to produce increased appetite and weight gain in long-term use (Paykel *et al.* 1973). Orthostatic hypotension, unsteadiness, falls, confusion, urinary retention, constipation, and cardiac effects can be particular problems in the elderly.

The dietary restrictions can be problematic with MAO inhibitors, as can such dose-related effects as orthostatic hypotension. Lithium causes weight gain, hypothyroidism, and fetal abnormalities which contraindicate it in pregnancy. Newer drugs might seem more attractive, but side effects can take a long time to become apparent after clinical introduction and those on long-term treatment will particularly be at risk: two drugs, nomifensine and zimeldine, have already been withdrawn for use because of side effects which only become manifest after considerable use. As a personal preference, I would still choose an older tricyclic at present in most cases for long-term use, since in maintenance doses, which tend to be submaximal, they produce relatively little in the way of problems.

It is important to note that the evidence for efficacy of long-term antidepressants depends mainly on relatively severe and recurrent depression. For such patients the benefits of antidepressants clearly outweigh the disadvantages. Milder depressions are common in the community and in general practice, and the lifetime incidence may be relatively high, and for these the advice may be different. The world would not thank psychopharmacologists for recommending long-term antidepressants for a substantial proportion of its population; nor would it accept their advice.

PREDICTORS OF SYMPTOM RETURN

Reliable predictors of symptom return would be of value in focusing therapeutic efforts. In short-term follow-up studies relapse has been found related to occurrence of stressful life events and absence of social support in the follow-up period (Paykel and Cooper 1992), to expressed emotion, particularly hostility, in key relatives (Vaughn and Leff 1976; Hooley *et al.* 1986), to neurotic symptom pattern (Paykel *et al.* 1974), older age of onset, and more previous episodes (Keller *et al.* 1983). Residual symptoms predicted higher relapse and recurrence rates over 1 year in one follow up of a trial of cognitive therapy versus antidepressant (Simons *et al.* 1986) and in one naturalistic follow up (Faravelli *et al.* 1986).

In longer-term follow-up studies higher recurrence rates have been found where there have been more previous episodes, and less consistently with endogenous or psychotic depression (Coryell and Winokur 1992).

Chronicity has been found associated with longer prior illness, personality neuroticism, higher family history, female sex (Weissman *et al.* 1978; Hirschfeld *et al.* 1986; Keller *et al.* 1986; Scott 1988; Scott *et al.* 1988).

There have been attempts to develop biological markers of incomplete recovery and risk of relapse. Efforts have been limited by the paucity of biological markers which satisfy the necessary criteria of being commonly abnormal in depression, clearly state rather than trait markers, and easily feasible for testing. A number of studies have followed up inpatients treated with tricyclic antidepressants, and found that persistent dexamethasone nonsuppression at the time of discharge predicted a greater risk of early relapse (Goldberg 1980; Greden *et al.* 1980; Holsboer *et al.* 1982; Targum 1984; Yerevanian *et al.* 1984; Schweitzer *et al.* 1987; Nemeroff and Evans 1988; Charles *et al.* 1989). One study of outpatients (Peselow *et al.* 1987) and two of patients treated with ECT (Coryell and Zimmerman 1983; Katona *et al.* 1987) have failed to find this. Coryell (1990) in a longer term 5-year follow up of major depressives found that nonsuppression on admission was associated with a greater incidence of suicide attempts regarded as psychologically serious, and also more development of mania. The dexamethasone suppression test is easy to carry out, but since no more than 50 per cent of depressives overall show nonsuppression even when depressed, it is of limited utility.

HIGH RECURRENCE RATES: IMPLICATIONS AND ALTERNATIVES

An emerging problem in all the antidepressant studies is that of high rates of relapse or recurrence after cessation of antidepressants. This problem was evident in the earlier short-term continuation studies, but was regarded as reflecting too early drug withdrawal prior to the natural remission of the disorder. However, a follow up of one of the early continuation studies, 1 year after 8 months continuation or earlier withdrawal (Weissman and Kasl 1976) found that the majority of subjects had sought further treatment during the year, suggesting a longer term problem.

Now the same finding emerges even after adequate continuation. All the studies in Table 13.2 employing adequate periods of continuation after initial treatment showed high recurrence rates on withdrawal to placebo: particularly noteworthy are the figures of around 80 per cent from the major studies of Frank *et al.* (1990) and Robinson *et al.* (1991). An earlier study by Bialos *et al.* (1982), in which patients receiving long-term amitriptyline for several years had medication withdrawn double blind, gave similar findings of 80 per cent recurrence within 15 weeks.

It is probable that the Frank and Robinson studies were biased towards high recurrence rates and drug efficacy by the requirement of previous episodes, good response to acute antidepressant and a symptom-free interval, prior to eligibility for the maintenance phase. Similarly the Bialos study is likely to have included a sample which had already done well on drugs and badly on earlier withdrawal.

Nevertheless, the findings suggest a problem. At best the effects of antidepressants appear disappointingly short term and noncurative. At worst the possibility arises that acute drug treatment may predispose to later relapse or recurrence. In essence, all the studies shown in Tables 13.1 and 13.2 which have adopted discontinuation designs have shown increased symptom return on withdrawal, compared with continuation of the drug. Drugs might produce long-term adaptation in CNS receptors rendering patients more susceptible to later depression. A withdrawal effect has been described for abrupt stoppage of high dose tricyclics administered for moderate periods (Kramer *et al.* 1961; Shatan 1966; Dilsaver and Greden 1984), but attributed plausibly to loss of atropinic side effects unmasking muscarinic supersensitivity (Dilsaver and Snider 1987), and reversed by atropine (Dilsaver *et al.* 1983). The atropinic effect does not occur with all similar antidepressants and is not believed to be related to the therapeutic effect. One pharmacological study (Ramana and Checkley 1990) has indicated that the uptake-inhibiting effect of the tricyclics is maintained in the longer term.

It is very difficult to demonstrate an increased vulnerability to depression after treatment because of uncertainty as to what the relapse rate might be without prior drug treatment and lack of a definitive biological test for vulnerability to depression. An increased rate of relapse early after withdrawal of lithium in bipolar patients has also been suggested (Christodoulou and Lykouras 1982; Mander 1986; Mander and Loudon 1988), although some evidence has also been advanced against such an early rebound phenomenon (Sashidharan and McGuire 1983).

I would not, at this stage, make a strong case for such an effect of antidepressants. I think it is more likely that the problem reflects the characteristics of recurrent patients who respond acutely to drugs and who are those selected for longer term studies. Detailed inspection of survival/recurrence curves where these are given in studies does not show a consistent pattern: in some studies symptom return does cluster soon after drug withdrawal, in others (for example, Montgomery *et al.* 1988; Frank *et al.* 1990) it clearly does not. The truth is probably that drug effects are short term and the disorder reverts to its sometimes poor natural history when they are withdrawn, but it would be unwise to reject confidently the sensitization explanation at this stage.

This possibility is also one argument against a suggestion that has sometimes been made that the focus for new drug evaluation might shift to long term efficacy. This would be making a drug which simply sensitized to subsequent depression, or a pure drug of dependence would come out well on the usual discontinuation design. New antidepressants do need evaluation in maintenance treatment for safety and efficacy but only as a subsidiary aim.

Another possibility to explain the findings is that the rate of drug withdrawal in the trials was too abrupt, precluding adaptation physiologically or psychologically to effects of coming off drug treatment. In most of the withdrawal studies the withdrawal appears to have been very rapid, and only rarely spread over as long as 3 weeks (Frank *et al.* 1990; Robinson *et al.* 1991) or 4 weeks (Paykel *et al.* 1975). In clinical practice slower withdrawal is often used.

Whatever the explanation, these findings raise the question of alternative nondrug approaches to prevention of relapse and recurrence. There is little evidence in favour of dynamic psychotherapy. In the study by Frank *et al.* (1990) the effect of interpersonal psychotherapy was small. In our continuation study where amitriptyline was effective in reducing the relapse rate, the early prototype for interpersonal therapy was not (Weissman *et al.* 1974; Paykel *et al.* 1975).

However, evidence has started to emerge that cognitive therapy may reduce rates of relapse and recurrence. Three follow-up studies over 6–24 months (Blackburn *et al.* 1986; Simons *et al.* 1986; S.D. Hollon, V.B. Yuason, M.J. Weiner, R.J. de Rubers, M.P. Evans, and M. Garvey, unpublished manuscript cited in Simons *et al.* 1986) of patients in acute treatment trials of tricyclic antidepressants and cognitive therapy showed high rates of relapse in the tricyclic treated groups, significantly lower relapse rates in the subjects who had received cognitive therapy. The differences were striking, with relapse rates of 66–78 per cent in drug groups, 12–25 per cent in cognitive therapy. Each study also included a group receiving both drug and cognitive therapy. In two of the three studies (Blackburn *et al.* 1986; S.D. Hollon, V.B. Yuason, M.J. Weiner, R.J. de Rubers, M.P. Evans, and M. Garvey, unpublished manuscript cited in Simons *et al.* 1986) these groups showed similar relapse rates to the cognitive therapy group. In the third study (Simons *et al.* 1986) the rate on combined treatment was higher than on cognitive therapy, raising the possibility of an adverse effect of the drug. A fourth study of imipramine, placebo, interpersonal psychotherapy, and cognitive therapy (Elkin *et al.* 1989) has not yet published follow-up findings, but is said to show some nonsignificant benefit for cognitive therapy on relapse (Beck, personal communication). A small fifth study (Miller *et al.* 1984) also showed a trend to lower relapse rates with cognitive therapy or social skills training than with standard treatment. However, two other follow-up studies have failed to show this benefit for cognitive therapy (Kovacs *et al.* 1981; Beck *et al.* 1985). The cognitive therapy studies have mainly been limited to outpatients or general practice patients, except for the Miller study.

The findings cannot be regarded as definitive. The follow up studies were naturalistic without fully controlled treatment, and cross-sectional at the end of the period rather than continuous. It is possible that different patients, less at risk of relapse, may have responded to cognitive therapy. What was mainly demonstrated was return to treatment rather than return of symptoms: cognitive therapy might provide an alternative self-managed coping mechanism. Drug continuation was only systematically controlled in one of the studies; systematic continuation would make a fairer comparison. It might nevertheless be argued that avoidance of relapse and recurrence without long-term drug taking is more desirable. It is possible that initial treatment with drugs raises an expectation of further drugs on relapse, while cognitive therapy presents an alternative coping strategy.

A good deal more work remains to be done, and particularly there is not a lot of evidence for cognitive therapy in more severe illnesses. Meanwhile all that can simply be said is that the story regarding long-term drug treatment does throw up some problems, and caution is still necessary before coming to enthusiastic recommendations for widespread long term maintenance.

CONCLUSIONS

The focus in depression is rapidly shifting from acute treatment, where for most patients the problems have been solved, or at least concern more the delivery of effective treatment than its efficacy, to longer term treatment, where it is becoming evident that the needs are considerable and the problems not so well solved. The evidence appears to be that in general, antidepressants which are effective in acute treatment are also effective in longer term treatment, but also, that their withdrawal, after shorter or longer term treatment, is associated with substantial rates of relapse and recurrence. They are not quite the definitive treatments which we once had hoped, once they are no longer being taken. Why this is so needs further exploration. Meanwhile, available trials provide good evidence that antidepressants lessen recurrence rates in recurrent unipolar depression. There is convincing evidence that antidepressants can help to prevent further attacks.

REFERENCES

ACNP (American College of Neuropsychopharmacology) (1990). Panel session: is dysthymia an affective disorder? Paper presented at *American college of neurosychopharmacology 25th annual meeting*, December 1990.

Angst, J. (1981). Clinical indications for a prophylactic treatment of depression. In *Depressive illness—biological psychopharmacological issues* (ed. J. Mendlewicz, A. Koppen, and H.M. van Praag), Advances in Biological Psychiatry, Vol. 7. S. Karger, Basal.

Baastrup, P.C., Poulsen, J.C., Schou, M., Thomsen, K., and Amdisen, A. (1970). Prophylactic lithium: double blind discontinuation in manic-depressive and recurrent-depressive disorders. *Lancet*, **ii**, 326–30.

Beck, A.T., Hollon, S.D., Young, J.E., Bedrosian, R.C., and Budenz, D. (1985). Treatment of depression with cognitive therapy and amitriptyline. *Archives of General Psychiatry*, **42**, 142–8.

Bialos, D., Giller, E., Jatlow, P., Docherty, J., and Harkness, L. (1982). Recurrence of depression after discontinuation of long-term amitriptyline treatment. *American Journal of Psychiatry*, **139**, 325–9.

Bjork, K. (1983). The efficacy of zimeldine in preventing depressive episodes in recurrent major depressive disorders—a double blind placebo-controlled study. *Acta Psychiatrica Scandinavica*, **68**, 182–9.

Blackburn, I.M., Eunson, K.M., and Bishop, S. (1986). A two year naturalistic follow up of depressed patients treated with cognitive therapy, pharmacotherapy and a combination of both. *Journal of Affective Disorders*, **10**, 67–75.

Charles, G.A., Schittecatte, M., Rush, A.J., Panzer, M., and Wilmotte, J. (1989). Persistent cortisol non-suppression after clinical recovery predicts symptomatic relapse in unipolar depression. *Journal of Affective Disorders*, **17**, 271–8.

Christodoulou, G.N. and Lykouras, E.P. (1982). Abrupt lithium discontinuation in manic-depressive patients. *Acta Psychiatrica Scandinavica*, **65**, 310–14.

Coppen, A. and Abou Saleh, M.T. (1983). Lithium in the prophylaxis of unipolar depression: a review. *Journal of the Royal Society of Medicine*, **76**, 297–301.

Coppen, A., Noguera, R., Bailey, J., Burns, B.H., Swani, M.S., Hare, E.H. *et al.* (1971). Prophylactic lithium in affective disorders. *Lancet*, **2**, 275–9.

Coppen, A., Montgomery, S., Gupta, R.K., and Bailey, J.E. (1976). A double-blind comparison of lithium carbonate and maprotiline in the prophylaxis of the affective disorders. *British Journal of Psychiatry*, **118**, 479–85.

Coppen, A., Ghose, K., Montgomery, S., Rao, V.A.R., Bailey, J., and Jorgensen, A. (1978a). Continuation therapy with amitryptyline in depression. *British Journal of Psychiatry*, **133**, 28–33.

Coppen, A., Ghose, K., Rao, V., Bailey, J., and Peet, M. (1978b). Mianserin and lithium in the prophylaxis of depression. *British Journal of Psychiatry*, **133**, 206–10.

Coppen, A., Abou-Saleh, M.T., Milln, P., Bailey, J., Metcalfe, M., Burns, B.H. *et al.* (1981). Lithium continuation therapy following electroconvulsive therapy. *British Journal of Psychiatry*, **139**, 284–7.

Coryell, W. (1990). DST abnormality as a predictor of course in major depression. *Journal of Affective Disorders*, **19**, 163–9.

Coryell, W. and Winokur, G. (1992). Course and outcome. In *Handbook of affective disorders* (ed. E.S. Paykel) (2nd edn). Churchill Livingstone, Edinburgh.

Coryell, W. and Zimmerman, M. (1983). The dexamethasone suppression test and ECT outcome: a six-month follow-up. *Biological Psychiatry*, **18**, 21–7.

Cundall, R.L., Brooks, P.W., and Murray, L.G. (1972). A controlled evaluation of lithium prophylaxis in affective disorders. *Psychological Medicine*, **2**, 308–11.

Davidson, J. and Raft, D. (1984). Use of phenelzine in continuation therapy. *Neuropsychobiology*, **11**, 191–4.

Dilsaver, S.C. and Greden, J.F. (1984). Antidepressant withdrawal phenomena. *Biological Psychiatry*, **19**, 237–56.

Dilsaver, S.C. and Snider, R.M. (1987). Amitriptyline produces dose-dependent supersensitivity of a central cholinergic mechanism. *Journal of Clinical Psychopharmacology*, **7**, 410–13.

Dilsaver, S.C., Feinberg, M., and Greden, J.F. (1983). Antidepressant withdrawal treated with anticholinergic agents. *American Journal of Psychiatry*, **140**, 244–51.

Doogan, D.P. and Caillard, V. (1993). Sertraline in the prevention of depression. *British Journal of Psychiatry*. (In press.)

Elkin, I., Shea, M.T., Watkins, J.T., Imber, S.D., Sotsky, S.M., Collins, J.F. *et al.* (1989). National Institute of Mental Health Treatment of Depression Collaborative Research Program: general effectiveness of treatment. *Archives of General Psychiatry*, **46**, 971–82.

Faravelli, C., Ambonetti, A., Pallanti, S., and Pazzagli, A. (1986). Depressive relapses and incomplete recovery from index episode. *American Journal of Psychiatry*, **143**, 888–91.

Fieve, R.R., Kumbaraci, T., and Dunner, D.L. (1976). Lithium prophylaxis of depression in bipolar I, bipolar II, and unipolar patients. *American Journal of Psychiatry*, **133**, 925–9.

Frank, E., Kupfer, D.J., Perel, J.M., Cornes, C., Jarrett, D.B., Maccinger, A.G. *et al.* (1990). Three year outcomes for maintenance therapies of recurrent depression. *Archives of General Psychiatry*, **47**, 1093–9.

Frank, E., Prien, R.F., Jarrett, R.B., Keller, M.B., Kupfer, D.J., Lavori, P.W. *et al.* (1991). Conceptualization and rationale for consensus definitions of terms in major depressive disorder: remission, recovery, relapse and recurrence. *Archives of General Psychiatry*, **48**, 851–5.

Georgotas, A. and McCue, R.E. (1989). Relapse of depressed patients after effective continuation therapy. *Journal of Affective Disorders*, **17**, 159–64.

Glen, A.I.M., Johnson, A.L., and Shepherd, M. (1984). Continuation therapy with lithium and amitriptyline in unipolar depressive illness: a randomized double-blind controlled trial. *Psychological Medicine*, **14**, 37–50.

Goldberg, I.K. (1980). Dexamethasone suppression test as indicator of safe withdrawal of antidepressant therapy. *Lancet*, **i**, 376.

Greden, J.F., Albala, A.A., Haskell, R.F., James, N.McI., Goodman, L., Steiner, M. *et al.* (1980). Normalization of dexamethasone suppression test: a laboratory index of recovery from endogenous depression. *Biological Psychiatry*, **15**, 449–58.

Guze, S.B. and Robins, E. (1970). Suicide and primary affective disorders. *British Journal of Psychiatry*, **117**, 437–8.

Harrison, W., Rabkin, J., Stewart, J.W., McGrath, P.J., Tricamo, E., and Quitkin, F. (1986). Phenelzine for chronic depression: a study of continuation treatment. *Journal of Clinical Psychiatry*, **47**, 346–9.

Hirschfeld, R.M.A., Klerman, G.L., Andreasen, N.C., Clayton, P.J., and Keller, M.B. (1986). Psycho-social predictors of chronicity in depressed patients. *British Journal of Psychiatry*, **148**, 648–54.

Holsboer, F., Liebl, R., and Hofschuster, E. (1982). Repeated dexamethasone suppression test during depressive illness. Normalisation of test result compared with clinical improvement. *Journal of Affective Disorders*, **4**, 93–101.

Hooley, J.M., Orley, J., and Teasdale, J.D. (1986). Levels of expressed emotion and relapse in depressed patients. *British Journal of Psychiatry*, **148**, 642–7.

Imlah, N.W., Ryan, E., and Harrington, J.A. (1965). The influence of antidepressant drugs on the response to ECT and on subsequent relapse rates. *Neuro-Psychopharmacology*, **4**, 438–42.

Jacob, M., Turner, L., Kupfer, D.J., Jarrett, D.B., Buzzinotti, E., and Bernstein, P. (1984). Attrition in maintenance therapy for current depression: a preliminary report. *Journal of Affective Disorders*, **6**, 181–9.

Johnstone, E.C., Owens, D.G.C., Lambert, M.T., Crow, T.J., Frith, C.D., and Done, D.J. (1990). Combination tricyclic antidepressant and lithium maintenance medication in unipolar and bipolar depressed patients. *Journal of Affective Disorders*, **20**, 225–33.

Kane, J.M., Quitkin, F.M., Rifkin, A., Ramos-Lorenzi, J.R., Nayak, D.P., and Howard, A. (1982). Lithium carbonate and imipramine in the prophylaxis of unipolar and bipolar II illness. *Archives of General Psychiatry*, **39**, 1065–9.

Katona, C.L.E., Aldridge, C.R., Roth, M., and Hyde, J. (1987). The dexamethasone suppression test and prediction of outcome in patients receiving ECT. *British Journal of Psychiatry*, **150**, 315–18.

Keller, M.B. and Shapiro, R.W. (1982). 'Double depression': superimposition of acute depressive episodes on chronic depressive disorders. *American Journal of Psychiatry*, **139**, 438–42.

Keller, M.B., Shapiro, R.W., Lavori, P.W., and Wolfe, N. (1982*a*). Recovery in major depressive disorder: analysis with the life table and regression models. *Archives of General Psychiatry*, **39**, 905–10.

Keller, M.B., Shapiro, R.W., Lavori, P.W., and Wolfe, N. (1982*b*). Relapse in major depressive disorder: analysis with the life table. *Archives of General Psychiatry*, **39**, 911–15.

Keller, M.B., Lavori, P.W., Lewis, C.E., and Klerman, G.L. (1983). Predictors of relapse in major depressive disorder. *Journal of the American Medical Association*, **250**, 3299–304.

Keller, M.B., Lavori, W., Rice, J., Coryell, W., and Hirschfeld, R.M.A. (1986). The persistent risk of chronicity in recurrent episodes of nonbipolar major depressive disorder: a prospective follow-up. *American Journal of Psychiatry*, **143**, 24–8.

Kiloh, L.G., Andrews, G., and Neilson, M. (1988). The long-term outcome of depressive illness. *British Journal of Psychiatry*, **153**, 752–7.

Klerman, G.L., DiMascio, A., Weissman, M., Prusoff, B., and Paykel, E.S. (1974). Treatment of depression by drugs and psychotherapy. *American Journal of Psychiatry*, **131**, 186–91.

Kovacs, M., Rush, A.J., Beck, A.T., and Hollon, S.D. (1981). Depressed outpatients treated with cognitive therapy or pharmacotherapy. *Archives of General Psychiatry*, **38**, 33–41.

Kramer, J.C., Klein, D.F., and Fink, M. (1961). Withdrawal symptoms following discontinuation of imipramine therapy. *American Journal of Psychiatry*, **118**, 549–50.

Lavori, P.W., Keller, M.B., and Klerman, G.L. (1984). Relapse in affective disorders: a reanalysis of the literature using life table methods. *Journal of Psychiatric Research*, **18**, 13–25.

Lee, A.S. and Murray, R.M. (1988). The long-term outcome of Maudsley depressives. *British Journal of Psychiatry*, **153**, 741–51.

Lehmann, H.E., Fenton, F.R., Deutsch, M., Feldman, S., and Engelsmann, F. (1988). An 11-year follow-up study of 110 depressed patients. *Acta Psychiatrica Scandinavica*, **78**, 57–65.

Mander, A.J. (1986). Is there a lithium withdrawal syndrome. *British Journal of Psychiatry*, **149**, 498–501.

Mander, A.J. and Loudon, J.B. (1988). Rapid recurrence of mania following abrupt discontinuation of lithium. *Lancet*, **2**, 15–17.

Miller, I.W., Norman, W.H., and Keitner, G.I. (1984). Cognitive-behavioral treatment of depressed inpatients: six- and twelve-month follow-up. *American Journal of Psychiatry*, **146**, 1274–9.

Mindham, R.H., Howland, C., and Shepherd, M. (1973). An evaluation of continuation therapy with tricyclic antidepressants in depressive illness. *Psychological Medicine*, **3**, 5–17.

Montgomery, S.A., Dufour, H., Brion, S., *et al.* (1988). The prophylactic efficacy of fluoxetine in unipolar depression. *British Journal of Psychiatry*, **153** (Suppl. 3), 69–76.

Nemeroff, C.B. and Evans, D.L. (1988). Correlation between the dexamethasone suppression test in depressed patients and clinical response. *American Journal of Psychiatry*, **141**, 247–9.

NIMH/NIH Consensus Development Conference Statement (1985). Mood disorders: pharmacologic prevention of recurrence. Consensus Development Panel. *American Journal of Psychiatry*, **142**, 469–76.

Paykel, E.S. (1989). Treatment of depression: the relevance of research for clinical practice. *British Journal of Psychiatry*, **155**, 754–63.

Paykel, E.S. and Cooper, Z. (1992). Life events and social support. In *Handbook of affective disorders* (ed. E.S. Paykel). Churchill Livingstone, Edinburgh.

Paykel, E.S. and Griffith, J.H. (1983). *Community psychiatric nursing for neurotic patients: the Springfield controlled trials*. Research Monographs in Nursing Series. Royal College of Nursing, London.

Paykel, E.S. and Tanner, K. (1976). Life events, depressive relapse and maintenance treatment. *Psychological Medicine*, **6**, 481–5.

Paykel, E.S., Mueller, P.S., and de la Vergne, P.M. (1973). Amitriptyline, weight gain and carbohydrate craving: a side effect. *British Journal of Psychiatry* m **123**, 501–7.

Paykel, E.S., Klerman, G.L., and Prusoff, B.A. (1974). Prognosis of depression and the endogenous-neurotic distinction. *Psychological Medicine*, **4**, 47–64.

Paykel, E.S., DiMascio, A., Haskell, D., and Prusoff, B.A. (1975). Effects of maintenance amitriptyline and psychotherapy on symptoms of depression. *Psychological Medicine*, **5**, 67–77.

Paykel, E.S., DiMascio, A., Klerman, G.L., Prusoff, B.A., and Weissman, M.M. (1976). Maintenance therapy of depression. *Pharmakopsychiatrie Neuro-Psychopharmacologie*, **9**, 127–36.

Peselow, E.D., Baxter, N., Rieve, R.R., Barouche, F., (1987). The dexamethasone suppression test as a monitor of clinical recovery. *American Journal of Psychiatry*, **144**, 30–5.

Prien, R.F. (1992). Maintenance treatment. In *Handbook of affective disorders* (ed. E.S. Paykel) (2nd edn). Churchill Livingstone, Edinburgh.

Prien, R.F. and Kupfer, D.J. (1986). Continuation drug therapy for major depressive episodes: how long should it be maintained? *American Journal of Psychiatry*, **143**, 18–23.

Prien, R.F., Klett, C.H., and Caffey, E.M. (1973a). Lithium carbonate and imipramine in prevention of affective episodes. *Archives of General Psychiatry*, **29**, 420–5.

Prien, R.F., Caffey, E.M., and Klett, C.H. (1973b). Prophylactic efficacy of lithium carbonate in manic-depressive illness. *Archives of General Psychiatry*, **28**, 337–41.

Prien, R.F., Kupfer, D.J., Mansky, P.A., Small, J.G., Tuason, V.B., Voss, C.B. *et al.* (1984). Drug therapy in the prevention of recurrences in unipolar and bipolar affective disorders. *Archives of General Psychiatry*, **41**, 1096–104.

Ramana, R. and Checkley, S.A. (1990). Biological effects of very long term antidepressant treatment. *International Review of Psychiatry*, **2**, 229–37.

Robinson, D.S., Lerfald, S.C., Bennett, B., Laux, D., Devereaux, E., Kayser, A. *et al.* (1991). Continuation and maintenance treatment of major depression with monoamine oxidase inhibitor phenelzine: a double-blind placebo-controlled discontinuation study. *Psychopharmacology Bulletin*, **27**, 31–9.

Rouillon, F., Phillips, R., Serrurier, D., Ansart, E., and Gérard, (1989). Rechutes de dépression unipolaire et efficacité de la maprotiline. *L'Encéphale*, **15**, 527–34.

Sashidharan, S.P. and McGuire, R.J. (1983). Recurrence of affective illness after withdrawal of long-term lithium treatment. *Acta Psychiatrica Scandinavica*, **68**, 126–33.

Schweitzer, I., Maguire, K.P., Gee, A.H., Tiller, J.W.G., Biddle, N., and Davies, B. (1987). Prediction of outcome in depressed patients by weekly monitoring with the dexamethasone suppression test. *British Journal of Psychiatry*, **151**, 780–4.

Scott, J. (1988). Chronic depression. *British Journal of Psychiatry*, **153**, 287–97.

Scott, J., Barker, W.A., and Eccleston, D. (1988). The Newcastle Chronic Depression Study: patient characteristics and factors associated with chronicity. *British Journal of Psychiatry*, **152**, 28–33.

Seager, C.P. and Bird, R.L. (1962). Imipramine with electrical treatment in depression. *Journal of Mental Science*, **108**, 704–7.

Shapiro, D.R., Quitkin, F.M., and Fleiss, J.L. (1989). Response to maintenance therapy in bipolar illness: effect of index episode. *Archives of General Psychiatry*, **46**, 401–5.

Shatan, C. (1966). Withdrawal symptoms after abrupt termination of imipramine. *Canadian Psychiatric Association Journal*, **11**(S), 150–8.

Simons, A.D., Murphy, G.E., Levine, J.L., and Wetzel, R.D. (1986). Cognitive therapy and pharmacotherapy of depression: sustained improvement over one year. *Archives of General Psychiatry*, **43**, 43–50.

Stein, M.K., Rickels, K., and Weisse, C.C. (1980). Maintenance therapy with amitriptyline: a controlled trial. *American Journal of Psychiatry*, **137**, 370–1.

Targum, S.D. (1984). Persistent neuroendocrine dysregulation in major depressive disorder: a marker for early relapse. *Biological Psychiatry*, **19**, 305–17.

Tyrer, P. (1984). Clinical effects of abrupt withdrawal from tricyclic antidepressants and monoamine oxidase inhibitors after long-term treatment. *Journal of Affective Disorders*, **6**, 1–7.

Vaughn, C.E. and Leff, J.P. (1976). The influence of family and social factors on the course of psychiatric illness: a comparison of schizophrenic and depressed neurotic patients. *British Journal of Psychiatry*, **129**, 125–37.

Weissman, M.M. and Kasl, S.V. (1976). Help-seeking in depressed outpatients following maintenance therapy. *British Journal of Psychiatry*, **129**, 252–60.

Weissman, M.M., Prusoff, B.A., and Klerman, G.L. (1978). Personality in the prediction of long term outcome of depression. *American Journal of Psychiatry*, **135**, 797–800.

Weissman, M.M., Klerman, G.L., Paykel, E.S., Prusoff, B.A., and Hanson, B.A. (1974). Treatment effects on the social adjustment of depressed outpatients. *Archives of General Psychiatry*, **30**, 771–8.

Yerevanian, B., Privitera, M., Milanese, E., Sagi, I., and Russotto, J. (1984). The dexamethasone test during major depressive episodes. *Biological Psychiatry*, **19**, 407–12.

14

New antidepressants—a look to the future

TIMOTHY H. CORN

INTRODUCTION

It has become customary to refer to antidepressant drugs as either first, second, or increasingly, and some would say over optimistically, as third generation. Looking at a list of the 26 antidepressants currently available in the UK, one of the best saturated markets, we can identify the grand-parents of this family: iproniazid introduced in 1958 and imipramine introduced the following year in 1959. This prototype monoamine oxidase inhibitor and monoamine reuptake inhibitor not only spawned a large number of similarly acting drugs, but also the monoamine theory of depression first formulated by Schildkraut in 1965 and still, with some modification, the mainstay of our understanding of the psychopharmacology of depression and of drug development in this area. We may indeed be currently testing third generation antidepressants but their grandparents are in most cases very easy to identify and still in clinical use. A review of Pharma Projects to October 1992 identifies a total of 128 drugs in development as antidepressants. The majority of these can be divided into four clearly defined groups: cleaned-up imipramine-like drugs, drugs whose primary action is on serotonergic systems, drugs whose primary action is on noradrenergic systems, and drugs which inhibit the action of the enzyme monoamineoxidase.

NOVEL ANTIDEPRESSANTS IN DEVELOPMENT OR RECENTLY INTRODUCED

Cleaned-up imipramine-like drugs

The primary objectives in the development of such drugs are as follows:

1. To reduce atropinic activity and hence related side effects.
2. To reduce sedative effects by reduction of interaction with muscarinic, histaminic, and α_1-adrenoceptor systems.

3. To reduce direct myocardial effects and hence increase safety in overdose.

This development pathway has not been much exploited by the pharmaceutical industry but lofepramine has achieved a moderate success in the clinic and does indeed seem to be safer in overdose than its predecessors, a prescription of lofepramine being ten times less likely to be associated with a fatal overdose than a prescription for an averagely safe antidepressant (Cassidy and Henry 1987).

Drugs acting on serotonergic systems

Serotonin reuptake inhibitors

The selective inhibitors of 5-HT reuptake have been the most successful group of antidepressants to emerge from the second generation of antidepressants. Table 14.1 lists those recently marketed or currently in development. Their selectivity for 5-HT uptake in animals has been clearly demonstrated (Benfield and Ward 1986) and their neuroendocrine effects such as the prolactin response to tryptophan (Price *et al.* 1989) combined with their typical, carcinoid-like side effect profile in man suggests that their effects in man do not differ significantly from those demonstrated preclinically. Despite suggestions that these drugs are rather less efficacious than conventional tricyclic drugs, for example the metanalyses performed by Bech (1990), fluoxetine is now the biggest selling antidepressant worldwide in cash terms.

The concerns, expressed particularly in the USA, that fluoxetine in particular can cause or enhance suicidal behaviour seem now to have been firmly laid to rest by the recently published metanalysis of all controlled trials of fluoxetine, including some 1765 patients, which fails to show that fluoxetine is associated with an increased risk of suicidal acts or emergence of substantial suicidal thoughts in depressed patients (Beasley *et al.* 1991).

TABLE 14.1. *Selective inhibitors of 5-HT uptake*

Fluoxetine	Fluvoxamine
Paroxetine	Sertraline
Citalopram	Cericlamine
(CGS 10686)	(Zimeldine)
Clovoxamine	Femoxetine
(Ifoxetine)	LY214281
SL81.0385	LY227942
VUFB15432	LY233708
Wy27587	Litoxetine
Nefazodone	Tiflucarbine

The majority of 5-HT uptake inhibitors in development would appear to offer little hope of advantage over the prototype drugs. However, nefazodone, which combines $5\text{-}HT_2$ receptor blockade with 5-HT reuptake inhibition, may be an exception to this generalization. In a trial reported by Mendels *et al.* (1991) 240 patients were randomized to 6 weeks treatment with low-dose nefazodone (150–300 mg per day), high dose nefazodone (300–600 mg per day) or placebo. Statistically significant differences were apparent for the high-dose group on HAM-D total, on retardation and anxiety subfactors, and on the CGI for improvement and severity.

Interestingly the time course of the HAM-D changes, again in the intent-to-treat (last observation carried forward) population clearly showed significant differences between placebo and high-dose nefazodone as early as 2 weeks post-initiation of treatment and turning to adverse effects it was observed that nefazodone shows a reasonably typical 5-HT uptake inhibitor profile but shows a significant degree of sedation. Given the putative role of $5\text{-}HT_2$ receptor antagonism in the treatment of psychosis and this evidence of some sedative effect it is tempting to speculate that nefazodone may be acting as a $5\text{-}HT_2$ antagonist in man and may prove to have a role in the more severe psychotic depressive illnesses which currently require combination therapy with an antidepressant and a neuroleptic.

The subclassification of 5-HT receptors is proceeding at such a pace that it is, in many instances, completely out stripping our understanding of the systems in which they operate. Checkley (1990) has pointed out that there are two ways in which drug developers can proceed in this situation. The first is to follow the simple logic:

1. 5-HT is involved in mood.

2. Drug X is a selective agonist/antagonist at a 5-HT receptor subtype.

3. Therefore test drug X in anxiety and depression.

The second is to attempt to tie the neurobiology of the novel receptor subtype to a known hypothesis about depression, such as the monoamine theory of depression, and proceed to develop appropriate agonists/antagonists at that receptor subtype.

$5\text{-}HT_{1A}$ receptor agonist drugs

$5\text{-}HT_{1A}$ agonists cause down-regulation of presynaptic $5\text{-}HT_{1A}$ receptors (Kennett *et al.* 1987), an effect which they share with antidepressants and ECT, both of which reduce the hypothermic response to challenge with the $5\text{-}HT_{1A}$ agonist 8-OH-DPAT (Goodwin *et al.* 1985). Hence buspirone, already marketed in anxiety, is being developed in depression, while ipsapirone and gepirone are both in Phase III development as antidepressants. In addition, zalospirone a $5\text{-}HT_{1A}$ partial agonist having a markedly antagonist profile in that it does not produce a typical 5-HT syndrome is said to be in Phase II development, and WY50324, which is a partial 5-HT_{1A} agonist of moderate efficacy and which binds to 5-HT_2 receptors with moderate affinity, is said to be in Phase I development (Muth *et al.* 1991).

Published clinical data for these compounds are, however, rather unimpressive. Buspirone, ipsapirone, and gepirone all appear to have some efficacy in depression (Robinson *et al.* 1990) although all published studies have significant methodological problems. In addition, both ipsapirone and gepirone have a poor side-effect profile and a limited therapeutic ratio, being associated with a high incidence of dysphoric dizziness at doses only marginally above those said to be efficacious in depression.

Hence it seems likely that these drugs would be difficult to use safely in clinical practice and have little advantage over existing therapy. Since all three of these compounds (and tandospirone) are converted metabolically to 1-(2-pyrimidinyl) piperazine (1-PP) (Caccia *et al.* 1986; Bianchi *et al.* 1988; Fischette *et al.* 1990) an α_2 antagonist (Rimele *et al.* 1987) with, in theory, antidepressant effects in its own right, the question of whether activation of 5-HT_{1A} receptors can lead to antidepressant activity cannot be answered at present (Fuller 1991).

5-HT₂ receptor antagonist drugs

A similar argument leads to the hypothesis that 5-HT_2 antagonists may have a role in the treatment of depression, in that both tricyclic antidepressants (Peroutka and Snyder 1980) and 5-HT_2 antagonists such as ritanserin (Leysen *et al.* 1986) cause down regulation of 5-HT_2 receptors. Ritanserin is currently in Phase III clinical development in both depression and schizophrenia.

Few clinical data have been published but efficacy in depression seems to be confined to mild to moderate depression and DSM-III-R Dysthymia (Aguglia *et al.* 1991) an indication which may be seen as pseudospecific. Whether or not this efficacy in depression is purely a function of 5HT_2 receptor blockade has recently been cast into doubt by the recognition that most 5-HT_2 receptor antagonists are also potent antagonists at 5-HT_{1C} receptors which in view of the marked structural similarity between the receptors revealed by cloning (Julius *et al.* 1990) is perhaps not surprising.

5-HT₁D receptor antagonist drugs

The final strategy involving 5-HT is based on the blockade of the presynaptic 5-HT autoreceptor. Such blockade should promote serotonergic transmission in exactly the same way that α_2 antagonists promote noradrenergic transmission. Terminal 5-HT autoreceptors mediating inhibition of 5-HT release belong to the 5-HT_1 receptor class and in rat have been identified as 5-HT_{1B} receptors. However little or no 5-HT_{1B} receptor binding can be demonstrated in man while 5-HT$_{1D}$ binding can be observed. Although pharmacologically distinct, 5-HT$_{1B}$ and 5-HT$_{1D}$ sites show similarities suggestive of a single ancestral gene and show similar distribution within the brains of the species in which they are present. Thus it can be argued that for antidepressant efficacy in man 5-HT$_{1D}$ receptor antagonism would be an appropriate drug development strategy (Hoyer and Middlemiss 1989) but as yet no compounds of this class appear to have found their way into the clinic.

Drugs acting on noradrenergic systems

Noradrenaline reuptake inhibitors
The sequential refinement of inhibitors of NA uptake has led through desipramine and maprotiline to oxaprotiline. This compound exists as a racemic mixture, the noradrenaline uptake inhibition residing solely in the (+) or S enantiomer (Waldmeier *et al.* 1977; Checkley *et al.* 1985). However it is the (−) or R enantiomer, levoprotiline, which is currently awaiting registration in Germany and which is currently in Phase III clinical trials in the US, and the rest of Europe.

The pharmacology of levoprotiline can be summarized as follows:

1. No effect on NA uptake.
2. Chronic treatment associated with adaptive changes at:
   (a) α_1 adrenoceptors
   (b) D-2 receptors
   (c) 5-HT receptors.
3. No effect on melatonin secretion in normal subjects.

Clinical data on levoprotiline has been rather slow in the publication and as yet no adequate placebo-controlled studies have been reported. However those studies which have been reported suggest that both the racemic oxaprotiline (Roffman *et al.* 1982) and enantiomerically pure levoprotiline (Delini-Stula *et al.* 1983) are effective antidepressants. This opens up a number of theoretical questions about the mechanism of action of R and S oxaprotiline many of which could be answered by a clinical study comparing the two isomers with placebo. This unique experiment in clinical psychopharmacology has yet to be performed but could have an impact on the development of the psychopharmacology of depression comparable to that of Crow's study of the structural isomers of flupenthixol in schizophrenia.

Beta receptor agonist drugs
It has been argued (Sulser 1984) that the down regulation of postsynaptic beta adrenoceptors might be a common pathway through which nora-

TABLE 14.2. α_2-*adrenoceptor anta-gonists*

Idazoxan	Fluparoxan
Atipamezole	Mirtazapine
Setiptiline	MK912
CH38083	Napamezole
Org 9768	RX821002
SC46264	

drenergic antidepressants exert their clinical effect. This down regulation could similarly be achieved by a centrally acting beta adrenoceptor agonist drug acting primarily at the cortically distributed beta-1 receptor (Minneman *et al.* 1979) rather than the cerebellar beta-2 receptor. As yet this drug development strategy has not been fully exploited, although flerobuterol is reported to be in early Phase II development.

α_2 adrenoceptor antagonist drugs

Antagonism of the presynaptic α_2 adrenoceptor which exerts inhibitory control over the release of noradrenaline (Langer 1981) offers a strategy for the elevation of intra synaptic noradrenaline and hence, an anti-depressant effect. A number of α_2 antagonists have been investigated in depression including those listed in Table 14.2.

Fluparoxan is an orally active α_2 antagonist which shows good selectivity for the α_2 receptor in *in vitro* studies having a selectivity ratio of greater than 2000 compared with 22 for the prototype antagonist yohimbine. In addition, fluparoxan shows extremely low affinity for a wide range of other receptor types. *In vivo* α_2 antagonism has been demonstrated both by the antagonism of UK 14304 induced sedation in the dog and by the inhibition of clonidine-induced growth hormone release in man. Early clinical trials in DSM-III-R Major Depression in which fluparoxan was compared to amitriptyline showed that fluparoxan was a safe and well-tolerated agent and suggested efficacy comparable to that of the standard tricyclic drug. As a result a series of placebo-controlled trials were undertaken in which several fixed doses of fluparoxan were compared with each other, and with placebo and tricyclic controls. These trials have confirmed the fluparoxan is a safe and well-tolerated drug but have demonstrated that at these safe and tolerable doses, the antidepressant activity of fluparoxan is limited; being somewhat greater than placebo but consistently less than imipramine.

Phosphodiesterase inhibitor drugs

It would clearly be theoretically possible to potentiate the function of the noradrenergic synapse by modifying second messenger mechanisms. Activation of central beta adrenoceptors leads to the production of second messenger cyclic AMP which in its turn is inactivated by the enzyme phosphodiesterase. Logically therefore blockage of this enzyme should have an antidepressant effect. Again this is a relatively unexplored route of antidepressant drug development with Rolipram the only example of its class. This compound is reported to be in late Phase III clinical development.

Clinical trials of rolipram versus amitriptyline have shown that the drug is well tolerated and, in general, has comparable efficacy with the standard agent (Eckmann *et al*. 1988). However Scott *et al*. (1991) in a trial comparing rolipram with amitriptyline in an inpatient group of depressed patients have shown very clearly, at least in hospitalized patients, that the efficacy of rolipram is not as great as that of standard tricyclic agents: amitriptyline being significantly superior to rolipram on all efficacy measures employed.

Reversible inhibitors of monoamine oxidase A

The monoamine oxidase inhibitors have a long history in psychiatric practice, iproniazid-induced change in mood being observed for the first time in 1956. Their clinical use has always been compromised by the cheese effect but recently introduced reversible inhibitors of monoamine oxidase A offer the prospect of antidepressant efficacy without tyramine-induced hypertension. A number of such compounds (listed in Table 14.3) are in late clinical research or have been recently launched.

To what extent have these expectations been fulfilled? As is usual little placebo-controlled evidence of efficacy has been published for many of these compounds. Moclobemide however has been launched in several countries and has been shown to be effective in a placebo-controlled comparison with imipramine in major depression (Versiani *et al*. 1989). Analysis of this data according to depressive subtype fails to demonstrate any marked advantage in nonendogenous depression and a review of side effects shows considerable advantage over standard tricyclic drugs.

Data for brofaromine and other agents in this class support the impression that these are efficacious antidepressants, but concerns have been raised about the claim that these agents are totally free of cheese effect (Corn 1990). This concern stems from the observation that the half-life for recovery of enzyme activity is considerably longer than the pharmacokinetic half-life of the inhibitor which implies some deviation from total reversibility as does the failure of attempts to reverse the inhibition of MAO by dilution

TABLE 14.3. *Reversible* MAOI *anti-depressants*

Moclobemide	Brofaromine
Bazinaprine	Toloxatone
Fezolamine	MD370503
RS8359	Sibutramine
BW137OU87	

or dialysis in tissues taken from animals pretreated with either moclobemide or brofaromine (Waldemeier *et al.* 1983; Waldemeier 1985). Pharmaco-epidemiological studies will have the final word on this matter.

BEYOND IMIPRAMINE AND IPRONIAZID

Despite the undoubted success of drug developers in refining these proto-type agents there are still many reasons to conclude that depressed patients are poorly served by current therapy. 30 per cent of patients with major depression fail to respond to available pharmacotherapy while patients with psychotic depression still require additional therapy with neuroleptic agents bringing further risk of adverse reactions. Despite the undoubted advances in antidepressant therapy patients are still subject to levels of adverse side effects that would not be tolerated in other areas of medicine and still find that the time to onset of effect is too long.

Arguably these residual problems affecting all currently available anti-depressants are the legacy of imipramine and iproniazid and no major advance in the pharmacotherapy of depression will be made until truly novel agents come forward. At present three areas of activity could pro-vide such a truly novel agent.

Dopamine

Several lines of evidence suggest that drugs acting through dopaminergic mechanisms may be antidepressant. In preclinical studies mesolimbic dopaminergic function is enhanced by electroconvulsive stimulation (Green and Deakin 1980) and by both tricyclic and atypical antidepressants (for review see Willner 1990). In clinical studies both directly acting dopamine agonists such as bromocriptine and piribedil (Post *et al.* 1978; Theohar *et al.* 1981) and indirectly acting compounds (such as nomifensine and bupro-pion) have been shown to have some antidepressant effects, in the case of the indirectly acting compounds comparable to that of standard agents.

Paradoxically low doses of neuroleptic drugs have been shown to have antidepressant effects (Robertson and Trimble 1982) and higher doses of neuroleptics in combination with tricyclic drugs or ECT are the mainstay in the treatment of psychotic depression. Clearly then there is a need to investigate the role of dopaminergic mechanisms in depression further using novel drug probes and the compound GBR 12909, a dopamine reuptake inhibitor, is interesting in this regard as it appears to have limited antidepressant activity in placebo-controlled clinical trials (Danion *et al.* 1991).

GABA

Evidence from several sources suggests the involvement of GABA in the pathogenesis of depressive illness. Depressed patients have low levels of GABA in both CSF and plasma, and have low levels of glutamic acid decarboxylase in frontal cortex.

All antidepressant drugs and ECT have the ability to increase $GABA_B$ binding in frontal cortex (Lloyd *et al.* 1989) suggesting that the elevation of cortical $GABA_B$ binding might be the final common pathway of anti-depressant treatment and indicating possible drug development targets. The GABA acting anticonvulsant drugs progabide and fengabine have indeed been studied in depression and, in comparative studies, appear to be effective however, as yet no placebo-controlled studies have been published. Other anticonvulsants such as carbamazepine and valproate have also been studied in depression and appear to have some effect although usually less marked than standard tricyclic agents (Calabrese *et al.* 1990).

Inositol-1-phosphate inhibitors

Until relatively recently the mechanism of action of lithium, a mainstay in the treatment of affective disorders, has been totally obscure. The discovery by Berridge and colleagues in 1982 that this ion inhibits the activity of the enzyme inositol-1-phosphatase has opened up the possibility of an enzyme target in the treatment of affective disorder.

The enzyme inositol-1-phosphatase participates in the phosphatidyl-inositol (PI) cycle second messenger system and catalyses the removal of the final phosphate from inositol triphosphate to produce inositol which can then combine with cytidine nucleotide derivative CDP-DG to form phosphatidylinositol. The supply of phosphatidylinositol is rate limiting in receptor activated PI turnover and hence blockade of the enzyme inositol-1-phosphatase by lithium could be expected to down regulate receptor-mediated PI breakdown and second messenger signal generation. As blockade of this enzyme occurs at concentrations of lithium within its normal therapeutic range it is now possible to formulate an explanatory hypothesis for the mechanism of action of lithium which implies the possibility of nonlithium, and possibly less toxic, treatments for affective disorders.

CONCLUSIONS

More than 30 years after the introduction of the prototypical antidepressants imipramine and iproniazid, the pharmacological treatment of depressive illness is still dominated by the same drugs or refinements of the same drugs. The depressed patient today can expect to be treated with a safer drug, both in terms of minor side effects and safety in overdose, than the patient 30 years ago but he or she cannot expect better absolute efficacy, faster response, or greater chance of response. There can be few areas of pharmacology and therapeutics where so little advance has been made.

What will the future bring for the pharmacological treatment of depression? The immediate future seems set to be dominated by the serotonin reuptake inhibitor antidepressants although the reversible inhibitors of monoaminoxide A may capture a significant share of the market. Looking into the more distant future there seems little cause for much optimism. A breakthrough in the pharmacological treatment of depression could be stimulated by one of two eventualities: the emergence of an entirely new understanding of the aetiology or pathogenesis of depression or the serendipitous finding of efficacy in an entirely novel class of pharmacological agent.

The convergent and constrained nature of the clinical development of new agents within the pharmaceutical industry using highly selected patient groups makes the chance finding of efficacy in depression for a drug being developed for a different clinical target highly unlikely. We therefore rely on the emergence of new biological understandings of depression to liberate us from the hegemony of imipramine, iproniazid, and their successors.

These new biological understandings of depression are a long time coming but may eventually emerge in the greater understanding of second messenger systems in particular those systems modified by lithium, a treatment also introduced many years ago which may only now be providing the biological clues needed for real progress to be made.

REFERENCES

Aguglia, E., Bersani, G., Bressa, G., De Majo, D., Giordano, P.L., Kemali, D. *et al.* (1991). Ritanserin in dysthymia. A multicentre double-blind trials vs. placebo. *Biological Psychiatry*, **29**, 328S.

Beasley, C.M., Jr, Dornseif, B.E., Bosomworth, J.C., Sayler, M.E., Rampey, A.H., Jr, Heiligenstein, J.H. *et al.* (1991). Fluoxetine and suicide: a meta-analysis of controlled trials of treatment for depression. *British Medical Journal*, **303**, 685–92.

Bech, P. (1990). A meta analysis of the antidepressant properties of serotonin reuptake inhibitors. *International Review of Psychiatry*, **2**, 207–11.

Benfield, P. and Ward, A. (1986). Fluvoxamine: a review of its pharmacodynamic and pharmacokinetic properties and therapeutic efficacy in depressive illness. *Drugs*, **32**, 313–34.

Berridge, M.J., Downes, C.P., and Hanley, M.R. (1982). Lithium amplifies agonist-dependent phosphatidylinositol responses in brain and salivary glands. *Biochemical Journal*, **206**, 587–95.

Bianchi, G., Caccia, S., Vedova, F.D., and Garattini, S. (1988). The adrenoceptor antagonist activity of ipsapirone and gepirone is mediated by their common metabolite 1-(2-pyrimidinyl)-piperazine(PmP). *European Journal of Pharmacology*, **151**, 365–71.

Caccia, S., Conti, I., Vigano, G., and Garattini, S. (1986). 1-(2-Pryimidinyl)-piperazine as active metabolite of buspirone in man and rat. *Pharmacology*, **33**, 46–51.

Calabrese, J.R., Gustavo, A., and Delucchi, G.A. (1990). The spectrum of efficacy of valproate in 55 patients with rapid-cycling bipolar disorder. *American Journal of Psychiatry*, **147**, 431–4.

Cassidy, S. and Henry, J. (1987). Fatal toxicity of antidepressant drugs in overdose. *British Medical Journal*, **295**, 1021–4.

Checkley, S.A. (1990). Criteria for assessing novel modes of antidepressant drug action in man. *International Review of Psychiatry*, **2**, 193–205.

Checkley, S.A., Thompson, C., Burton, S., Franey, C., and Arendt, J. (1985). Clinical studies of the effect of (+) and (−) oxaprotiline upon noradrenaline uptake. *Psychopharmacology*, **87**, 117–18.

Corn, T.H. (1990). Reversible inhibitors of monoamine oxidase A: antidepressants without cheese effect? *International Review of Psychiatry*, **2**, 187–92.

Danion, J.M., Montgomery, S.A., Snel, S., and Skrumsager, B.K. (1991). A placebo controlled study with GBR 12909 a dopamine uptake inhibitor. *Biological Psychiatry*, **29**, 635S.

Delini-Stula, A., Vassout, A., Hausen, K., Bittiger, H., Buech, U., and Olpe, H.-R. (1983). Oxaprotiline and its enantiomers; do they open up new avenues in the research in the mode of action of antidepressants. *Frontiers in neuropsychiatric research* (ed. E. Usdin, M. Goldstein, A. Friedhoff, and A. Georgotas), pp. 121–34. Mcmillan, London.

Eckmann, F., Fichte, K., Meya, U., and Sastrey-Hernandez, M. (1988). Rolipram in major depression: results of a double-blind comparative study with amitriptyline. *Current Therapeutic Research*, **43**, 291–5.

Fischette, C.T., Delgado, P.L., Seibyl, J., Krystal, J.H., Heninger, G.R., and Charney, D.S. (1990). Neuroendocrine effects of tandospirone (SM-3997) in healthy male subjects. *Society for Neuroscience Abstracts*, **16**, 914.

Fuller, R.W. (1991). Serotonin in depression and related disorders. *Journal of Clinical Psychiatry*, **52** (Suppl.), 52–7.

Goodwin, G.M., De Souze, R.J., and Green, A.R. (1985). Presynaptic serotonin receptor-mediated response in mice attenuated by antidepressant drugs and electroconvulsive shock. *Nature*, **317**, 531–3.

Green, A.R. and Deakin, J.F.W. (1980). Brain noradrenaline depletion prevents ECS-induced enhancement of serotonin- and dopamine-mediated behaviour. *Nature*, **285**, 232–3.

Hoyer, D. and Middlemiss, D. (1989). Species differences in the pharmacology of terminal 5HT autoreceptors in mammalian brain. *Trends in Pharmacological Sciences*, **10**, 130–2.

Julius, D., Huang, K.N., Livelli, T.J., Axel, R., and Jessell, M. (1990). The 5-HT$_2$ receptor defines a family of structurally distinct but functionally conserved serotonin receptors. *Proceedings of the National Academy of Science, USA*, **87**, 928–32.

Kennett, G.A., Marcou, M., Dourish, C.T., and Curzon, G. (1987). Single administration of 5-HT$_{1A}$ agonists decreases 5-HT$_{1A}$ presynaptic but not post-synaptic receptor mediated responses: relationship to antidepressant-like actions. *European Journal of Pharmacology*, **138**, 53–60.

Langer, S.Z. (1981). Presynaptic regulation of the release of catecholamines. *Pharmacology Reviews*, **32**, 337–62.

Leysen, J.E., Van Gompel, P., Gommeren, W., Woestenborghs, R., and Janssen, P.A.J. (1986). Down-regulation of serotonin-S2 receptor sites in rat brain by chronic treatment with the serotonin-S2 antagonists Ritanserin and Setoperone. *Psychopharmacology*, **88**, 434–44.

Lloyd, K.G., Zivkovic, D., Scatton, B., Morselli, P.L. and Bartholini, G. (1989). The GABAergic hypothesis of depression. *Progress in Neuropsychopharmacology and Biological Psychiatry*, **13**, 341–51.

Mendels, J., Reimherr, F., Roberts, D., Ecker, J., Marcus, R., Schwiderski, U., and Robinson, D. (1991). A double blind comparison of high dose nefazodone, low dose nefazodone and placebo in the treatment of depressed outpatients. *European Neuropsychopharmacology*, **1**, 451–2.

Minneman, K.P., Hegstrand, L.R., and Molinoff, P.B. (1979). Simultaneous determination of beta-1 and beta-2-adrenergic receptors in tissues containing both receptor subtypes. *Molecular Pharmacology*, **16**, 34–46.

Muth, E.A., Moyer, J.A., Haskins, J.T., Abou-Gharbia, M.A., Stephens, R.J., and Ward, T.J. (1991). Zalospirone, Wy-50, 324, and WAY-100,289: new serotonergic agents with psychotherapeutic potential. *European Neuropsychopharmacology*, **1**, 207–9.

Peroutka, S.J. and Snyder, S.H. (1980). Long-term antidepressant treatment decreases spiroperidol-labelled serotonin receptor binding. *Science*, **210**, 88–90.

Post, R.M., Gerner, R.H., Carman, J.S., Gillin, C., Jimerson, D.C., Goodwin, F.K. *et al.* (1978). Effects of a dopamine agonist piribedil in depressed patients: relationship of pretreatment homovanillic acid to antidepressant response. *Archives of General Psychiatry*, **35**, 609–15.

Price, L.H., Charney, D.S., Delgado, P.L., Anderson, G.M., and Heninger, G.R. (1989). Effects of despiramine and fluvoxamine treatment on the prolactin response to tryptophan. Serotonergic function and the mechanism of antidepressant action. *Archives of General Psychiatry*, **46**, 625–31.

Rimele, T.J., Henry, D.E., Lee, D.K.H., Geiger, G., Heaslip, R.J., and Grimes, D. (1987). Tissue-dependent α-adrenoceptor activity of buspirone and related compounds. *Journal of Pharmacology and Experimental Therapeutics*, **241**, 771–8.

Robertson, M.M. and Trimble, M.R. (1982). Major tranquillisers used as antidepressants. *Journal of Affective Disorders*, **4**, 173–93.

Robinson, D.S., Rickels, K., Feighner, J., Fabre, L.F., Jr, Gammans, R.E., Shrotriya, R.C. *et al.* (1990). Clinical effects of the 5-HT$_{1A}$ partial agonists in depression: a composite analysis of buspirone in the treatment of depression. *Journal of Clinical Psychopharmacology*, **10** (Suppl. 3), 67S–76S.

Roffmann, M., Gould, E.E., Brewer, S.J., Lau, H., Sachais, B., Dixon, R.B. *et al.* (1982). A double-blind comparison study of oxaprotiline with amitriptyline and placebo in moderate depression. *Current Therapeutic Research*, **32**, 247–56.

Schildkraut, J.J. (1965). The catecholamine hypothesis of affective disorders: a review of supporting evidence. *American Journal of Psychiatry*, **122**, 509–22.

Scott, A.I.F., Perini, A.F., Schering, P.A., and Whalley, L.J. (1991). In-patient major depression: is rolipram as effective as amitryptyline? *European Journal of Clinical Pharmacology*, **40**, 127–9.

Sulser, F. (1984). Regulation and function of noradrenaline receptor systems in brain. Psychopharmacological aspects. *Neuropharmacology*, **23**, 255–61.

Theohar, C., Fischer-Cornelssen, K., Akesson, H.O., Ansari, J., Gerlach, J., Harper, P. *et al.* (1981). Bromocriptine as antidepressant: double-blind comparative study with imipramine in psychogenic and endogenous depression. *Current Therapeutic Research*, **30**, 830–42.

Versiani, M., Oggero, U., Alterwain, P., Capponi, R., Dajas, F., Heinze-Martin, G. *et al.* (1989). A double-blind comparative trial of moclobemide v. imipramine and placebo in major depressive episodes. *British Journal of Psychiatry*, **155** (Suppl. 6), 72–7.

Waldemeier, P.C. (1985). On the reversibility of reversible MAO inhibitors. *Naunyn-Schmiedeberg's Archives of Pharmacology*, **329**, 305–10.

Waldemeier, P.C., Baumann, P.A., Wilhelm, M., Bernasconi, R., and Maitre, L. (1977). Selective inhibition of noradrenaline and serotonin uptake by C 49802 B- and CGP 6985A. *European Journal of Pharmacology*, **46**, 387–91.

Waldmeier, P.C., Felner, A.E., and Tipton, K.F. (1983). The monoamine oxidase inhibiting properties of CGP 11305 A. *European Journal of Pharmacology*, **94**, 20–6.

Willner, P. (1990). The role of slow changes in catecholamine receptor function in the action of antidepressant drugs. *International Review of Psychiatry*, **2**, 141–56.

Index